# Practice Questions
in Pure
Mathematics

# Practice Questions in Pure Mathematics

R.L. Smith and C. Denton

Edward Arnold

© C. Denton and R.L. Smith 1981

First published 1981
by Edward Arnold (Publishers) Ltd.
41 Bedford Square, London WC1B 3DQ

All rights reserved. No part of this publication may be reproduced, stored in a retrieval system, or transmitted in any form or by any means, electronic, mechanical, photocopying, recording or otherwise, without the prior permission of Edward Arnold (Publishers) Ltd.

**British Library Cataloguing in Publication Data**

Smith, R. L.
   Practice questions in pure mathematics.
   1. Mathematics – Problems, exercises, etc.
   I. Title     II. Denton, C.
   510.76     QA43

ISBN 0-7131-0434-1

Photo Typeset by Macmillan India Ltd. Bangalore

Printed and bound in Great Britain
at The Camelot Press Ltd, Southampton

# Preface

Our aim has been to provide a book of revision notes and exercises covering all the main topics in an 'A' level syllabus for the 'pure' section of a 'pure and applied' course. We have referred closely to the syllabi and past examinations of the J.M.B. and London boards but the book will be equally useful for the examinations of the other main boards which offer a traditional syllabus.

Each chapter is preceded by brief notes covering the essential material needed for the exercises. When a particular topic merits extra practice, parallel A and B exercises are included. The exercises should be tackled in the most convenient order and no attempt has been made to suggest what that order might be.

Engineering students and students following correspondence courses will find this a valuable reference and a source of appropriate exercises, as will those who are studying additional level mathematics. The latter should omit the miscellaneous exercises which are generally of a higher degree of difficulty.

It is with regret that I must report the death of my colleague Ron Smith before the completion of the book. He was a teacher of many years and many kinds of experience and his influence on the book's development was invaluable.

C.D.

# Contents

1 Algebraic Operations
   1.1 Notes and Formulae    1
   1.2 Indices and Logarithms    3
   1.3 Partial Fractions    4
   1.4 Elimination    5
   1.5 Proof by Induction    6
   1.6 Miscellaneous    6

2 Quadratic Equations and Functions
   2.1 Notes and Formulae    7
   2.2 Quadratic Equations and Quadratic Functions    8
   2.3 Symmetrical Functions of the Roots of a Quadratic    10
   2.4 Simultaneous Equations    11
   2.5 Miscellaneous    11

3 Permutations, Combinations and Probability
   3.1 Notes and Formulae    12
   3.2 Permutations    14
   3.3 Combinations    15
   3.4 Miscellaneous Permutations and Combinations    16
   3.5 Probability    17

4 Series
   4.1 Notes and Formulae    19
   4.2 Arithmetic Progressions    22
   4.3 Geometric Progressions    22
   4.4 The Binomial Theorem (for Positive Integral Indices)    23
   4.5 Other Series    24
   4.6 Miscellaneous (Finite Series)    25
   4.7 Infinite Geometric Series    26
   4.8 The Binomial Series    27
   4.9 Exponential and Logarithmic Series    27
   4.10 Taylor's and Maclaurin's Theorems    28
   4.11 Miscellaneous    29

5 Trigonometric Identities and Equations
   5.1 Notes and Formulae    31

|   |      |                                           |    |
|---|------|-------------------------------------------|----|
|   | 5.2  | Graphs and Ratios for any Angle           | 32 |
|   | 5.3  | Identities and Use of Basic Formulae      | 35 |
| 6 | Trigonometric Formulae                           |    |
|   | 6.1  | Notes and Formulae                        | 37 |
|   | 6.2  | Addition Formulae                         | 39 |
|   | 6.3  | Sums and Differences                      | 43 |
|   | 6.4  | Small Angles and Graphical Solutions      | 46 |
|   | 6.5  | Miscellaneous                             | 47 |
| 7 | Solution of Triangles                            |    |
|   | 7.1  | Notes and Formulae                        | 48 |
|   | 7.2  | Solution of Triangles and Miscellaneous Areas | 49 |
|   | 7.3  | Miscellaneous Problems with Triangles     | 50 |
|   | 7.4  | Miscellaneous Problems in Three Dimensions | 52 |
| 8 | Curves                                           |    |
|   | 8.1  | Notes and Formulae                        | 54 |
|   | 8.2  | Co-ordinates                              | 55 |
|   | 8.3  | Areas in Polar Co-ordinates               | 56 |
|   | 8.4  | Curve Sketching of Rational Functions     | 57 |
|   | 8.5  | Miscellaneous Curves                      | 57 |
| 9 | Inequalities                                     |    |
|   | 9.1  | Notes and Formulae                        | 58 |
|   | 9.2  | Basic Inequalities                        | 59 |
|   | 9.3  | Quadratic Inequalities                    | 60 |
|   | 9.4  | Miscellaneous                             | 61 |
| 10 | Co-ordinate Geometry                            |    |
|   | 10.1 | Notes and Formulae                        | 62 |
|   | 10.2 | The Straight Line                         | 66 |
|   | 10.3 | Loci                                      | 68 |
|   | 10.4 | The Circle                                | 70 |
|   | 10.5 | The Parabola                              | 72 |
|   | 10.6 | The Ellipse and Hyperbola                 | 74 |
|   | 10.7 | Miscellaneous                             | 76 |
| 11 | Determination of Laws                           |    |
|   | 11.1 | Notes and Formulae                        | 77 |
|   | 11.2 | Examples on Determination of Laws         | 78 |
| 12 | Complex Numbers                                 |    |
|   | 12.1 | Notes and Formulae                        | 80 |
|   | 12.2 | Basic Operations                          | 81 |
|   | 12.3 | Complex Roots                             | 82 |
|   | 12.4 | Geometrical Representation                | 83 |
|   | 12.5 | Equating Real and Imaginary Parts         | 84 |
|   | 12.6 | Miscellaneous                             | 84 |

| | | |
|---|---|---|
| **13** | **Differentiation** | |
| | 13.1 Notes and Formulae | 85 |
| | 13.2 Differentiation of Powers of $x$ | 87 |
| | 13.3 Function of a Function | 88 |
| | 13.4 Further Basic Differentiation | 88 |
| | 13.5 Products and Quotients | 89 |
| | 13.6 Further Differentiation Methods | 90 |
| | 13.7 Miscellaneous | 91 |
| **14** | **Applications of Differentiation** | |
| | 14.1 Notes and Formulae | 92 |
| | 14.2 Gradients, Tangents, and Normals | 93 |
| | 14.3 Maxima, Minima and Points of Inflexion | 94 |
| | 14.4 Maxima and Minima Problems | 94 |
| | 14.5 Small Increments and Connected Rates of Change | 95 |
| | 14.6 Distance, Velocity, and Acceleration | 96 |
| | 14.7 Miscellaneous | 97 |
| **15** | **Integration and Differential Equations** | |
| | 15.1 Notes and Formulae | 98 |
| | 15.2 Integration of Simple Functions | 100 |
| | 15.3 Integration Needing Manipulation, Including Partial Fractions | 101 |
| | 15.4 Integration by Substitution | 103 |
| | 15.5 Integration by Parts | 104 |
| | 15.6 Differential Equations | 105 |
| | 15.7 Miscellaneous | 106 |
| **16** | **Applications of Integration** | |
| | 16.1 Notes and Formulae | 107 |
| | 16.2 Areas | 108 |
| | 16.3 Volumes | 109 |
| | 16.4 Centroids and Centres of Gravity | 110 |
| | 16.5 Mean Values | 111 |
| | 16.6 Approximate Integration | 111 |
| | 16.7 Miscellaneous | 112 |
| **Answers** | | 113 |

# 1
# Algebraic Operations

## 1.1 Notes and Formulae

**Logarithms**

If $N = a^x$ then $x = \log_a N$. (It is important to be able to use this connection without hesitation. Remember a logarithm is an index.)

$\log A + \log B = \log AB$

$\log A - \log B = \log \dfrac{A}{B}$

$\log A^n = n \log A$

$\log \sqrt[n]{A} = \dfrac{1}{n} \log A$

$\log_a N = \dfrac{\log_b N}{\log_b a}$ (An important result.)

$\log_a b = \dfrac{1}{\log_b a}$  $\log_e 10 = 2.2036$  $\log_{10} e = 0.43429$

**Partial Fractions**

*Worked Examples*

(a) $\dfrac{5}{(2x-1)(x+2)} = \dfrac{A}{2x-1} + \dfrac{B}{x-2}$

$5 = A(x-2) + B(2x-1)$
Put $x = -2$:   $5 = -5B \Rightarrow B = -1$
Put $x = \frac{1}{2}$:   $5 = 2\frac{1}{2}A \Rightarrow A = 2$

(b) $\dfrac{9x+9}{(x-3)(x^2+9)} = \dfrac{A}{x-3} + \dfrac{Bx+C}{x^2+9}$
(Note treatment of 2nd degree denominator.)

$9x+9 = A(x^2+9) + (Bx+C)(x-3)$
Put $x = 3$:   $36 = 18A$   $\Rightarrow A = 2$
Put $x = 0$:   $9 = 9A - 3C$   $\Rightarrow C = 3$

Put $x = 1$: $\quad 18 = 10A - 2B - 2C \Rightarrow B = -2$

(c) $\dfrac{3x^2 - 20x + 34}{(x-3)^3} = \dfrac{A}{(x-3)^3} + \dfrac{B}{(x-3)^2} + \dfrac{C}{x-3}$

(Note treatment of repeated factor.)

$$3x^2 - 20x + 34 = A + B(x-3) + C(x-3)^2$$

Equating coefficients (an alternative method):

$$\begin{array}{lll} x^2: & 3 = C & \Rightarrow C = 3 \\ x: & -20 = B - 6C & \Rightarrow B = -2 \\ \text{Const:} & 34 = A - 3B + 9C & \Rightarrow A = 1 \end{array}$$

## Elimination

There is no set routine for the elimination of one variable from two equations; the student must use his initiative. Two common methods are illustrated below.

*Substitution*

Given $(x-1) = t(y-2)$ and $tx + y = 2$ to eliminate $t$:

$$t = \dfrac{x-1}{y-2} \quad \text{from the first equation,}$$

$$\dfrac{x-1}{y-2} x + y = 2 \quad \text{on substitution in the second equation}$$

or $x(x-1) + (y-2)^2 = 0$

*Use of Identities*

Given $\sin\theta = \tfrac{1}{2}(y-3)$ and $\cos\theta = \tfrac{1}{2}(x+1)$ to eliminate $\theta$:

$$\begin{array}{ll} \text{As} & \sin^2\theta + \cos^2\theta = 1 \\ & \tfrac{1}{4}(y-3)^2 + \tfrac{1}{4}(x+1)^2 = 1 \\ \text{or} & (x+1)^2 + (y-3)^2 = 4 \end{array}$$

## Induction

The basic method is to assume that the identity which has to be proved is true for $n$. Then add one more term (or step) to each side and show that the result (on the right hand side) when simplified shows that the identity is true for $(n+1)$. Finally show that the identity is true for the first possible value, usually $n = 1$. If this is true then the result holds for $n = 2$, $n = 3$, etc., i.e. for all positive integral values of $n$.

*Worked Example*

Prove that $\dfrac{1}{1.4} + \dfrac{1}{4.7} + \dfrac{1}{7.10} \ldots$ to $n$ terms $= \dfrac{n}{3n+1}$.

Assume the result is true for $n$.
Add the next term to both sides.

$$\frac{1}{1.4} + \frac{1}{4.7} + \frac{1}{7.10} + \ldots + \frac{1}{(3n+1)(3n+4)}$$

$$= \frac{n}{3n+1} + \frac{1}{(3n+1)(3n+4)}$$

$$= \frac{3n^2 + 4n + 1}{(3n+1)(3n+4)}$$

$$= \frac{n+1}{3n+4}$$

The right hand side is the same as that in the given identity but with $n$ replaced by $n+1$. Hence the result is true for $n+1$ if it is true for $n$.
If $n = 1$ both sides are equal to $\frac{1}{4}$; hence it is true for $n = 1$. Hence it is true for $n = 2, 3, 4$, etc.

## 1.2 Indices and Logarithms

**Exercise A**
1. Evaluate $5^5$, $7^4$, $2^{10}$, $(-\frac{1}{2})^7$, $(-\frac{1}{3})^5$, $10^0$.
2. Simplify $81^{1/2}$, $64^{1/3}$, $81^{1/4}$, $225^{1/2}$, $27^{2/3}$, $(-5)^0$.
3. Evaluate $10^{-1}$, $5^{-2}$, $3^{-4}$, $3^0$, $(\frac{1}{2})^{-5}$, $(\frac{1}{25})^{-1/2}$.
4. Simplify $216^{1/3}$, $64^{2/3}$, $36^0$, $81^{3/4}$, $32^{-4/5}$, $(\frac{25}{49})^{-1/2}$.
5. Using the approximations $\sqrt{2} = 1.414$, $\sqrt{3} = 1.732$, $\sqrt{5} = 2.236$, find approximate values for $\sqrt{8}$, $\sqrt{12}$, $\sqrt{20}$, $\sqrt{0.5}$, $\sqrt{0.2}$.
6. Evaluate $\log_2 8$, $\log_3 81$, $\log_4 16$, $\log_5 625$.
7. Simplify $\log_4 8$, $\log_3 \frac{1}{27}$, $\log_4 \frac{1}{32}$, $\log_5 0.008$, $\log_4 1$, $\log_9 27$, $\log_8 16$, $\log_9 \frac{1}{243}$.
8. Solve the following equations:
   (a) $2 \log_x 64 = 3$
   (b) $\log_x 2 + \log_x 32 = 2$
   (c) $\log_3 x + 3 \log_x 3 = 4$
   (d) $\frac{6}{\log_x 4} + \frac{6}{\log_4 x} = 13$
   (e) $\log_9 \left(\frac{x}{3}\right) = \log_9 x \log_x 3$
9. Solve the equations
   (a) $3 \log_2 x + \log_2 y = 10$, $4 \log_2 x - 2 \log_2 y = 0$
   (b) $3 \log_4 x + 2 \log_2 y = 12$, $\log_2 x - 2 \log_4 y = 1$
   (c) $4 \log_4 x - 3 \log_5 y = 1$, $\log_2 x - \log_5 y = 2$
   (d) $5x - y = 2$, $\log_x y = 1 + \log_x 4$
   (e) $\log y = 2 \log x + \log 2$, $\log (y - x - 2) = 0$
   (f) $\log_3 x - \log_3 y = 4$, $\log_2 (x - 33y) = 4$

3

## Exercise B

1. Using the approximations $\sqrt{2} = 1.414$, $\sqrt{3} = 1.732$, $\sqrt{6} = 2.449$, find approximate values for $\sqrt{32}$, $\sqrt{24}$, $\sqrt{27}$, $\sqrt{\frac{1}{3}}$, $\sqrt{\frac{1}{6}}$.
2. Evaluate $\log_4 64$, $\log_5 125$, $\log_9 81$, $\log_3 243$.
3. Simplify $\log_9 27$, $\log_2 \frac{1}{16}$, $\log_5 \frac{1}{25}$, $\log_2 0.125$.
4. Evaluate $\log_5 1$, $\log_4 \frac{1}{32}$, $\log_{16} 32$, $\log_9 \frac{1}{729}$.
5. Solve the following equations:
   (a) $2 \log_x 243 = 5$
   (b) $\log_{10} x + \log_{10} 4 = 2$
   (c) $\log_2 x \log_4 x = 4.5$
   (d) $16 \log_x 2 = \log_2 x$
   (e) $2 \log_4 x + 3 \log_x 4 = 7$
6. Solve the equations
   (a) $2 \log_3 x + \log_3 y = 5$, $\log_3 x - 2 \log_3 y = 5$
   (b) $2 \log_3 x + \log_2 y = 10$, $4 \log_3 x - 3 \log_2 y = 0$
   (c) $2 \log_5 x + 3 \log_4 y = 3$, $\log_5 x - \log_4 y = 4$
   (d) $2 \log y = \log 2 + \log x$, $2x + 4y - 5 = 0$
   (e) $2 \log x - \log y = \log 4$, $\log(x - y + 1) = 0$
   (f) $\log_4 x = \log_2 y - 2\frac{1}{2}$, $\log_2 x = \log_4 y + 1$

## 1.3 Partial Fractions

Separate each of the following expressions into partial fractions:

### Exercise A

1. $\dfrac{5x + 1}{(x - 1)(x + 2)}$
2. $\dfrac{x - 9}{x^2 - 8x + 15}$
3. $\dfrac{9x + 10}{(x - 2)(x + 5)}$
4. $\dfrac{16x - 11}{6x^2 - 5x - 6}$
5. $\dfrac{8x - 10}{4x^2 - 1}$
6. $\dfrac{1}{(2x - 3)(x + 1)}$
7. $\dfrac{x}{(2x - 1)(x + 4)}$
8. $\dfrac{2}{3x^2 - 8x - 3}$
9. $\dfrac{3x}{2x^2 + 3x - 5}$
10. $\dfrac{x^2 + 5x + 16}{(x + 1)(x + 3)(x - 2)}$
11. $\dfrac{9x - 10}{(2x^2 + 5)(2x - 1)}$
12. $\dfrac{13x - 5}{(x^2 + 5)(2x - 3)}$
13. $\dfrac{9x + 9}{(x + 2)^2 (x - 1)}$
14. $\dfrac{3x^2 - 20x + 34}{(x - 3)^3}$

### Exercise B

1. $\dfrac{9 - 2x}{(x - 2)(x + 3)}$
2. $\dfrac{x + 5}{x^2 + 3x + 2}$

3. $\dfrac{2x+33}{2x^2-19x+9}$      4. $\dfrac{x+35}{9x^2-25}$

5. $\dfrac{5x+5}{6x^2-x-1}$      6. $\dfrac{1}{(2x+5)(x-2)}$

7. $\dfrac{x}{(3x+2)(x-1)}$      8. $\dfrac{x}{6x^2-x-1}$

9. $\dfrac{4x^2+16x-14}{(x-3)(x+2)(x+4)}$      10. $\dfrac{13x-9}{(3x^2-1)(x+5)}$

11. $\dfrac{5x+1}{x^3-1}$      12. $\dfrac{1-23x}{(x^2+2x+3)(x-5)}$

13. $\dfrac{67-20x}{(x-3)^2(x+4)}$      14. $\dfrac{12x^2-14x+6}{(2x-1)^3}$

## 1.4 Elimination

### Exercise A
Eliminate $t$ (or $\theta$) in each of the following pairs of equations to obtain an equation connecting $x$ and $y$.

1. $x = a\cos\theta,\ y = a\sin\theta$      2. $x = t,\ y = \dfrac{1}{t}$

3. $x = a\sec\theta,\ y = b\tan\theta$      4. $x = at,\ y = t + \dfrac{1}{t}$

5. $x = a\sin\theta,\ y = a\sin 2\theta$      6. $x = 1+t,\ y = 3t^2 - 2t^3$

7. $x = a\cos^3\theta,\ y = a\sin^3\theta$      8. $x = \dfrac{1-t}{1+t},\ y = \dfrac{1-3t}{1+t}$

9. $x = a\cos\theta,\ y = a\cos 2\theta$      10. $x = 2\sin\theta + 3,\ y = 3\cos\theta - 2$

11. $x = \log z^2 + 3,\ y = \log\left(\dfrac{1}{z}\right)$      12. $x = a\sin\theta,\ y = b\sec\theta$

13. $x = t^2 - 1,\ y = t^4 - 1$      14. $x = t + \dfrac{1}{t},\ y = t - \dfrac{1}{t}$

### Exercise B
Eliminate $t$ or $\theta$ from each of the following pairs of equations:

1. $x = a\sec\theta,\ y = a\tan\theta$      2. $x = a\cos\theta,\ y = b\sin\theta$

3. $x = ct,\ y = \dfrac{c}{t}$      4. $x = at^2,\ y = bt$

5. $x = \cos 2\theta,\ y = \sin\theta$      6. $x = 2t+3,\ y = t-1$

7. $x = 1+\cos\theta,\ y = 1-\sin\theta$      8. $x = m-2,\ y = m^2 - 4$

9. $x = 2\operatorname{cosec}\theta,\ y = 2\cot\theta$      10. $x = 3\tan\theta - 1,\ y = 2\cot\theta - 3$

11. $\log x = 2t,\ \log y = t+1$      12. $x = p\sin\theta,\ y = q\tan\theta$

13. $x = t^3 - 1,\ y = t - 1$      14. $x = t + \dfrac{1}{t},\ y = t^2 + \dfrac{1}{t^2}$

## 1.5 Proof by Induction

Prove the following identities by induction:

**Exercise A**
1. $1 + 2 + 3 + \ldots + n = \frac{1}{2}n(n+1)$
2. $1 + 4 + 7 + \ldots$ to $n$ terms $= \frac{1}{2}n(3n-1)$
3. $3 + 9 + 27 + 81 + \ldots$ to $n$ terms $= \frac{3}{2}(3^n - 1)$
4. $a + ar + ar^2 + \ldots$ to $n$ terms $= \dfrac{a(r^n - 1)}{r - 1}$
5. $1^3 + 2^3 + 3^3 + \ldots + n^3 = \dfrac{n^2(n+1)^2}{4}$
6. $1.2 + 2.3 + 3.4 + \ldots$ to $n$ terms $= \frac{1}{3}n(n+1)(n+2)$
7. $1.1! + 2.2! + 3.3! + \ldots + n.n! = (n+1)! - 1$
8. $\dfrac{1}{1.2.3} + \dfrac{1}{2.3.4} + \dfrac{1}{3.4.5} + $ to $n$ terms $= \dfrac{n(n+3)}{4(n+1)(n+2)}$
9. $1.2.3 + 2.3.5 + 3.4.7 + \ldots + n(n+1)(2n+1) = \frac{1}{2}n(n+1)^2(n+2)$

**Exercise B**
1. $1 + 3 + 5 + \ldots + (2n-1) = n^2$
2. $3 + 7 + 11 + \ldots$ to $n$ terms $= n(2n+1)$
3. $1 + 2 + 4 + 8 + \ldots$ to $n$ terms $= 2^n - 1$
4. $a + (a+d) + (a+2d) + \ldots$ to $n$ terms $= \frac{1}{2}n\{2a + (n-1)d\}$
5. $1^2 + 2^2 + 3^2 + \ldots$ to $n$ terms $= \frac{1}{6}n(n+1)(2n+1)$
6. $1.2.3 + 2.3.4 + 3.4.5 + \ldots$ to $n$ terms $= \frac{1}{4}n(n+1)(n+2)(n+3)$
7. $\dfrac{1}{1.3} + \dfrac{1}{3.5} + \dfrac{1}{5.7} + \ldots$ to $n$ terms $= \dfrac{n}{2n+1}$
8. $1.2.3.4 + 2.3.4.5 + 3.4.5.6 + \ldots$ to $n$ terms $= \frac{1}{5}n(n+1)(n+2)(n+3)(n+4)$
9. $1.2^2 + 2.3^2 + 3.4^2 + \ldots$ to $n$ terms $= \frac{1}{12}n(n+1)(3n^2 + 11n + 10)$

## 1.6 Miscellaneous

1. Prove that $\log_4 x = \frac{1}{2} \log_2 x$ and $\log_8 p = \frac{1}{3} \log_2 p$.
2. Simplify (a) $\log_2 x + \log_4 x + \log_{16} x$  (b) $\log_2 x + \log_4 x \log_x 2$
3. Prove that $\log_a b \log_b c \log_c a = 1$.
4. Express $y = \dfrac{2x^2 - x - 4}{(2x+1)(x-1)}$ in partial fractions and hence

    (a) find an expression for $\dfrac{d^2y}{dx^2}$ and (b) evaluate $\displaystyle\int \dfrac{2x^2 - x - 4}{(2x+1)(x-1)} dx$.

5. Find all the values of $\dfrac{y}{x}$ which satisfy the equation

$$2 \log y = \log x + \log(4x + 3y)$$

6. Show, by induction or otherwise, that 3 is a factor of $n^3 + 3n^2 + 2n$ for all positive integral values of $n$.
7. If $u_1, u_2, u_3$ form a geometric series prove that $\log u_1, \log u_2, \log u_3$ form an arithmetic series.
8. From the equations $x + y + w = 1$, $x + z + w = 6$, and $x + 2y + 3z = 6$ eliminate $w$ and $z$ to obtain an equation in $x$ and $y$.
9. If $x^2 \cos^2 \theta - xy \sin \theta \cos \theta + 2y^2 \sin^2 \theta = 0$, and $x \sin \theta = y \cos \theta$, eliminate $\theta$ to obtain an equation in $x$ and $y$ only.
10. In a given plane there are $n$ straight lines of which no two are parallel and no three are concurrent. Prove by induction, or otherwise, that the number of regions into which the plane is divided is $\frac{1}{2}(n^2 + n + 2)$. (Hint: add one more line and find the number of extra regions formed.)

# 2

# Quadratic Equations and Functions

## 2.1 Notes and Formulae

If the roots of $ax^2 + bx + c = 0$ are $\alpha$ and $\beta$ then

$$\alpha + \beta = -\frac{b}{a} \qquad \alpha\beta = \frac{c}{a}$$

If the sum of the roots of a quadratic equation is $S$ and the product is $P$ then the equation is

$$x^2 - Sx + P = 0$$

**Useful identities**

$$\alpha^2 + \beta^2 = (\alpha + \beta)^2 - 2\alpha\beta$$
$$(\alpha - \beta)^2 = (\alpha + \beta)^2 - 4\alpha\beta$$
$$\alpha^3 + \beta^3 = (\alpha + \beta)\{(\alpha + \beta)^2 - 3\alpha\beta\}$$

If $ax^2 + bx + c = 0$, then

$$x = \frac{-b \pm \sqrt{b^2 - 4ac}}{2a}$$

(i) If $b^2 > 4ac$ the roots are real and different. The roots are rational if $b^2 - 4ac$ is a perfect square; otherwise they are irrational.

(ii) If $b^2 = 4ac$ the roots are real and equal.

7

(iii) If $b^2 < 4ac$ the roots are unreal.

For all real values of $x$ the expression $f(x) = ax^2 + bx + c$ has the same sign as $a$ except when the roots of $f(x) = 0$ are real and unequal and $x$ lies between them.

To find the range of possible values of $y$, where $y = \dfrac{ax^2 + bx + c}{px^2 + qx + r}$, rearrange the equation as a quadratic in $x$ and use conditions (i) and (ii) above.

## 2.2 Quadratic Equations and Quadratic Functions

### Exercise A

1. Solve the following equations, giving roots to 3 significant figures (if not exact) or listing them as unreal where appropriate.
   (a) $2x^2 + 7x - 4 = 0$ (b) $x^2 - 6x - 2 = 0$ (c) $x^2 - 2x + 6 = 0$
   (d) $5x^2 - 4x - 3 = 0$ (e) $4x^2 + 5x + 6 = 0$ (f) $3x^2 - 2x - 1 = 0$

2. Without solving the following equations state the nature of the roots (real and rational, real and irrational, real and equal, or unreal).
   (a) $3x^2 - 10x + 3 = 0$ (b) $x^2 - 14x + 49 = 0$ (c) $2x^2 - 3x - 4 = 0$
   (d) $x^2 + 3x + 5 = 0$ (e) $9x^2 - 6x + 1 = 0$ (f) $3x^2 - 2x - 3 = 0$
   (g) $4x^2 + 5x - 6 = 0$ (h) $2x^2 - 5x + 4 = 0$ (i) $4x^2 - 20x + 25 = 0$

3. Find the value of $a$ if
   (a) $x^2 - ax + 25 = 0$ has equal roots.
   (b) $x^2 + ax - 15 = 0$ has two integral roots.
   (c) $x^2 + ax + 6b^2 = 0$ has two roots which are integral multiples of $b$.
   (d) $6x^2 - 7x + a = 0$ has two roots which are reciprocals.

4. Write down the conditions that the roots of $ax^2 + bx + c = 0$ are
   (a) equal in magnitude but opposite in sign. (b) reciprocals.

5. Find the range of values of $a$ for which
   (a) $x^2 + ax + 25 = 0$ has real roots.
   (b) $x^2 + ax + 16 = 0$ has no real roots.
   (c) $x^2 + ax - 1 = 0$ has real roots.

6. What is the value of $b$ if $x^2 + (b-2)x + (b^2 - b - 2) = 0$
   (a) has one, and only one, zero root?
   (b) has two zero roots?
   (c) has roots of which one is twice the other?
   (d) has two equal roots?

7. What is the value of $c$ if $cx^2 - 2(c+1)x + (c+3) = 0$
   (a) has two equal roots?
   (b) has one, and only one, zero root?

8. If $x$ is real find the range of possible values of
   (a) $2x^2 + 3x - 1$ (b) $\dfrac{x^2 - 2}{2x + 1}$ (c) $\dfrac{3x^2 + 2}{x^2 + 4x}$ (d) $\dfrac{5x^2 - 2}{4 - x}$

9. For what range of possible values of $x$ will the following functions of $x$ be positive?
   (a) $(3x + 4)(x - 2)$ (b) $2x^2 - x - 1$ (c) $3x - 2x^2 + 5$

(d) $3(a-2)x^2 - 3ax + 6$ $(a > 4)$  (e) $\dfrac{2x-1}{(3+x)(2-x)}$

(f) $\dfrac{2x^2 + x - 10}{3x^2 + 8x + 4}$

10. Sketch the graphs of the following functions:
    (a) $y = (x-1)(x+2)$   (b) $y = (x-3)(x-5)$
    (c) $y = \dfrac{x-1}{x+2}$   (d) $y = \dfrac{1}{(x-1)(x+2)}$

## Exercise B

1. Without solving the following equations state the nature of the roots (real and rational, real and irrational, real and equal, or unreal).
    (a) $2x^2 + 5x + 3 = 0$   (b) $2x^2 + 6x + 3 = 0$   (c) $3x^2 - 2x + 1 = 0$
    (d) $3x^2 - 4x + 1 = 0$   (e) $4x^2 + x + 1 = 0$   (f) $4x^2 + 2x + 3 = 0$
    (g) $4x^2 + 3x + 2 = 0$   (h) $12x^2 - 17x - 5 = 0$
    (i) $15x^2 - 26x - 8 = 0$

2. Find the value of $p$ if
    (a) $x^2 + px - 21 = 0$ has two integral roots.
    (b) $x^2 + px - 6 = 0$ has one root which is 2.
    (c) $x^2 + 10x + p = 0$ has two equal roots.
    (d) $x^2 + px + p = 0$ has two equal roots.

3. Find the condition that the roots of $px^2 + qx + r = 0$ are
    (a) equal.   (b) add up to zero.

4. Find the range of values of $p$ for which
    (a) $x^2 + px + 25 = 0$ has no real roots.
    (b) $x^2 + 2x + p = 0$ has real roots.
    (c) $px^2 + 2x + 3 = 0$ has real roots.

5. Find the value of $s$ if $x^2 - 2(s-3)x + (s^2 - 4s - 5) = 0$
    (a) has one, and only one, zero root.
    (b) has two roots numerically equal but opposite in sign.
    (c) has two equal roots.

6. Find the value of $r$ if $x^2 + 2(r+3)x + (r^2 + 2r - 5) = 0$
    (a) has two equal roots.
    (b) has one, and only one, zero root.

7. If $x$ is real find the possible range of values of
    (a) $4x - x^2 - 3$   (b) $\dfrac{4x^2 + 2}{x^2 - 2x}$   (c) $\dfrac{x^2 - 1}{2x^2 - 2x}$   (d) $\dfrac{5x^2 + 2}{4 + x}$

8. For what range of values of $x$ will the following functions be positive?
    (a) $(2x - 3)(3x + 2)$   (b) $3 - 2x^2 - x$   (c) $\dfrac{x-4}{x+1}$

(d) $\dfrac{(3x+1)(x-5)}{(x+1)}$  (e) $\dfrac{2x^2-11x+12}{4x^2-17x+40}$

9. Sketch the graphs of the following functions:

(a) $y = (x-2)(x+3)$   (b) $y = \dfrac{x-2}{x+3}$   (c) $y = \dfrac{1}{(x-2)(x+3)}$

## 2.3 Symmetrical Functions of the Roots of a Quadratic

**Exercise A**

1. The roots of the equation $x^2 - 3x + 1 = 0$ are $\alpha, \beta$. Find the values of
   (a) $\alpha^2 + \beta^2$   (b) $(\alpha - \beta)^2$   (c) $\alpha^3 + \beta^3$   (d) $\alpha^3 - \beta^3$

2. The roots of the equation $2x^2 + 3x + 3 = 0$ are $\alpha, \beta$. Find the value of
   (a) $\dfrac{1}{\alpha} + \dfrac{1}{\beta}$   (b) $\dfrac{1}{\alpha^2} + \dfrac{1}{\beta^2}$   (c) $\alpha^2\beta + \alpha\beta^2$   (d) $\dfrac{\alpha^2}{\beta} + \dfrac{\beta^2}{\alpha}$

3. Given that the roots of the equation $x^2 + ax + a^2 - 1 = 0$ are $r$ and $s$ ($r \geq s$), find the value of   (a) $r^2 + rs + s^2$   (b) $\dfrac{r}{s^2} + \dfrac{s}{r^2}$   (c) $r^4 - s^4$

4. If the roots of $x^2 + 2x + 3 = 0$ are $\alpha$ and $\beta$, find the equation whose roots are
   (a) $2\alpha, 2\beta$   (b) $\alpha^2, \beta^2$   (c) $\dfrac{1}{\alpha}, \dfrac{1}{\beta}$   (d) $\alpha^3, \beta^3$

5. If the roots of the equation $2x^2 - x + 3 = 0$ are $\alpha$ and $\beta$, find the equation whose roots are
   (a) $\dfrac{\alpha}{\beta}, \dfrac{\beta}{\alpha}$   (b) $\dfrac{1}{\alpha^2}, \dfrac{1}{\beta^2}$   (c) $\alpha + 1, \beta + 1$   (d) $2\alpha + \beta, \alpha + 2\beta$

6. If one root of $ax^2 + bx + c = 0$ is twice the other show that $2b^2 = 9ac$.

**Exercise B**

1. The roots of $x^2 + 3x - 4 = 0$ are $\alpha, \beta$. Find the value of
   (a) $\alpha^2 + \beta^2$   (b) $\alpha^2 - \alpha\beta + \beta^2$   (c) $\alpha^3 + \beta^3$   (d) $\dfrac{\alpha}{\beta} + \dfrac{\beta}{\alpha}$

2. The roots of $3x^2 - 2x - 6 = 0$ are $\alpha, \beta$ ($\alpha > \beta$). Find the value of
   (a) $\alpha^3 - \beta^3$   (b) $\dfrac{1}{\alpha^3} + \dfrac{1}{\beta^3}$   (c) $\alpha^3\beta + \alpha\beta^3$   (d) $\dfrac{\alpha^3}{\beta} + \dfrac{\beta^3}{\alpha}$

3. The roots of $x^2 + p^2x + p = 0$ are $\gamma, \delta$ ($\gamma > \delta$). Find the value of
   (a) $\gamma^2 - \gamma\delta + \delta^2$   (b) $\gamma^2 - \delta^2$   (c) $\gamma^4 - \delta^4$

4. The roots of $x^2 - 3x - 1 = 0$ are $\alpha, \beta$. Find the equation with roots
   (a) $-\alpha, -\beta$   (b) $\dfrac{1}{\alpha^2}, \dfrac{1}{\beta^2}$   (c) $1 + \dfrac{1}{\alpha}, 1 + \dfrac{1}{\beta}$

5. Given that the equation $4x^2 - 2x + 3 = 0$ has roots $\alpha, \beta$, find the equation with roots
   (a) $\alpha^2\beta, \alpha\beta^2$ (b) $\dfrac{\alpha^2}{\beta}, \dfrac{\beta^2}{\alpha}$ (c) $\dfrac{\alpha+1}{\beta}, \dfrac{\beta+1}{\alpha}$ (d) $\dfrac{\alpha^3}{\beta}, \dfrac{\beta^3}{\alpha}$
6. One of the roots of the equation $x^2 + ax + b^3 = 0$ is the square of the other. Prove that $a + b + b^2 = 0$.

## 2.4 Simultaneous Equations

### Exercise A
1. Solve the following pairs of equations:
   (a) $3x - y = 1$, $2x^2 + y^2 = 6$   (b) $x + 2y = 5$, $x^2 + xy = 12$
   (c) $3x - 2y = 5$, $x^2 - y^2 + xy = 11$   (d) $x - 3y = 1$, $x^2 - 2xy + 2y^2 = 2$
2. Find the points in which the line $y = 2x + 3$ meets the curve $y = x^2 - 2x + 6$.
3. Show that $x - 2y + 5 = 0$ is a tangent to the curve $y^2 - 4y - 2x + 6 = 0$ and find the point of contact.
4. Show that $3x = 2y - 7$ does not intersect $y^2 = 5x - 2$ in real points.
5. If $5x = 7y + 4$ intersects $y^2 = 2x - 1$ at $A$ and $B$ find the mid-point of $AB$.
6. Find the condition that $y = 2x^2 - a$ cuts $x^2 + y^2 = 1$ in
   (a) 4   (b) 3   (c) 2   (d) 1   (e) no real points.
7. Find the condition that $2y = 7x + 2$ is a tangent to $y = ax^2 + bx + c$.

### Exercise B
1. Solve the following pairs of equations:
   (a) $y = 2x - 6$, $x^2 - y^2 = 12$   (b) $y = 2x - 3$, $x^2 + xy - y^2 = 5$
   (c) $y = 3x - 9$, $x^2 + y^2 - 2x - 4y - 20 = 0$
   (d) $2x + 3y = 5$, $2x^2 + y^2 + 4x = 8$
2. Find the points in which the curve $y^2 = 8x$ is intersected by the line $y + 8 = 2x$.
3. Show that the line $y = 4x - 1$ is a tangent to the curve $2y = x^2 + 4x + 2$ and find its point of contact.
4. Show that the curve $x^2 + 4y^2 = 4$ and the line $y = 2x + 5$ do not intersect in real points.
5. Find the mid-point of the chord formed by the intersection of the line $y = 2x - 2$ and the curve $y^2 - 4y = 2x - 6$.
6. Find the mid-point of the chord formed by the intersection of the line $y = x - 2$ and the curve $9x^2 + 16y^2 = 144$.
7. Find the value of $a$ such that $y = x + a$ is a tangent to $x^2 + 4y^2 = 4$.

## 2.5 Miscellaneous

1. Find the larger root of $x^2 - 12x + 0.05 = 0$ correct to 3 significant figures. Why is it not possible to find the other root to the same degree of accuracy when four-figure tables are used? Use the product of the roots to find the smaller root.

2. Show that if $\alpha$ and $\beta$ are the roots of $ax^2 + bx + c = 0$ then $\dfrac{1}{\alpha}$ and $\dfrac{1}{\beta}$ are the roots of $cx^2 + bx + a = 0$.
3. The roots of $x^2 + px + p - 1 = 0$ are $\alpha$ and $\beta$. The roots of $x^2 + (p+2)x + p + 2 = 0$ are $\alpha + 1$ and $\beta + 1$. Find $p$ and hence solve the equations.
4. Show that the roots of $(x-p)(x-q) = r^2$ are real. ($p$, $q$, $r$ are all real constants.)
5. Find $P$ if $x^2 + 3(2-P)x + 4(2P-3) = 0$ has equal roots.
6. Show that the roots of $x^2 + 2Px + P^2 - Q^2 = 0$ are real. ($P$, $Q$ are real constants.)
7. Show that the roots of $(P-Q)x^2 + (2P+Q)x + P + 2Q = 0$ are real. ($P$, $Q$ are real constants.)
8. What values of $k$ make $x^2 + (k-1)x + (k-2)$ positive for all real values of $x$.
9. If the roots of $x^2 + 2x + p = 0$ are $\alpha$ and $\beta$, and the roots of $px^2 - 4x + q = 0$ are $(\alpha + \beta)^2$ and $(\alpha - \beta)^2$ find $p$, $q$, $\alpha$, $\beta$.
10. $f(x) = ax^2 + 2x + 3$ and $g(x) = x^2 + 2x + 3b$. $f(x) = 0$ if $x = \gamma$ or $\delta$, and $g(x) = 0$ if $x = \gamma - 3\delta$ or $\delta - 3\gamma$. Find $a$ and $b$.

# 3

# Permutations, Combinations and Probability

## 3.1 Notes and Formulae

**Permutations and Combinations**

If an operation can be performed in '$p$' ways and a second operation can be performed in '$q$' ways, then the number of ways of performing the first followed by the second is '$pq$'. This result can be extended to more than two operations by induction.

$n!$ ($n$ factorial) is calculated as $n(n-1)(n-2)\ldots 3.2.1$
e.g. $4! = 4.3.2.1 = 24$

A **permutation** of a number of elements of a set is any *arrangement* of some or all of the set.

A **combination** is any *selection* of some or all of the set.

The number of permutations of '$n$' different elements taken '$r$' at a time is written

$$^nP_r$$

and is equal to
$$\frac{n!}{(n-r)!}$$

If we define 0! as equal to 1, this result is true for the special case when $n = r$. Hence the number of permutations of $n$ elements taken all together $= n!$

The number of combinations of $n$ elements taken $r$ at a time (written $^nC_r$ or $\binom{n}{r}$) is
$$\frac{n!}{r!(n-r)!}$$

The number of permutations of $n$ elements when all are included and $p$ of them are identical is
$$\frac{n!}{p!}$$

The total number of ways of making a selection from $n$ objects when each can be either included or excluded is $2^n - 1$. (If at least one object is to be taken.)

## Probability

If an event can happen in '$n$' equally likely ways and of these '$r$' are regarded as 'successful' then the probability of a 'successful' result is $\frac{r}{n}$; e.g. the probability of drawing an ace from an ordinary pack of cards is $\frac{4}{52}$ or $\frac{1}{13}$.

If A represents the occurrence of a certain event and $\overline{A}$ represents its non-occurrence then $p(\overline{A}) = 1 - p(A)$, where $p(A)$ is the probability that the event will occur.

If B is another event then $A \cup B$ means the event that A occurs or B occurs or both. $A \cap B$ means that A *and* B occur.

$$p(A \cup B) = p(A) + p(B) - p(A \cap B).$$

If A and B are mutually exclusive $p(A \cap B) = 0$ and so
$$p(A \cup B) = p(A) + p(B)$$

$p(A|B)$ means the conditional probability that A will occur when B is known to have occurred.
$$p(A|B) = \frac{p(A \cap B)}{p(B)}$$
and $p(A \cap B) = p(B) p(A|B)$

If event B is *independent* of event A then
$$p(B) = p(B|A) \quad \text{and} \quad p(A \cap B) = p(A) p(B)$$

When there are a finite number of outcomes counting techniques are useful for calculating probabilities

$$p(A) = \frac{\text{number of ways A can happen}}{\text{total number of outcomes}}$$

$$p(A|B) = \frac{\text{number of ways A and B can happen together}}{\text{number of ways B can happen}}$$

Taking a selection at *random* means that the selection is made in such a way that the actual probability of each event is the same as the theoretical probability, no constraints being applied to any possible outcome.

If the probability of an event is p, the odds that it will occur are $p:(1-p)$.

## 3.2 Permutations

1. Compute (a) 6! (b) 7! (c) 5! (d) 10!
2. Compute (a) $^6P_2$ (b) $^8P_5$ (c) $^8P_7$ (d) $^{100}P_3$ (e) $^{100}P_2$.
3. List the permutations of the letters A, B, C, D
   (a) taking 1 at a time. (b) taking 2 at a time. (c) taking 3 at a time.
4. In how many ways can the letters of MASTER be arranged?
5. How many different arrangements are possible for six pipes in a pipe rack?
6. Three cards are taken from an *Ace, King, Queen, Jack and Ten*. How many different permutations can be formed?
7. Thirty people each buy one raffle ticket. If there are three prizes, in how many different ways can the prizes be won?
8. How many (a) 3 letter arrangements, (b) 4 letter arrangements, can be made from the letters of the word CANISTER if each letter can be used only once?
9. Six children run a race. In how many ways can the first three places be filled?
10. Fifteen beads of different colours are in a bag. In how many ways can four beads be chosen and put in a row?
11. How many permutations are there of the letters of the following words if all the letters are to be included? (a) STEAMER (b) FEEBLE (c) CALCULUS (d) SISTERS (e) CANAAN (f) REVIVER (g) MISSISSIPPI
12. A school has six mathematics teachers. There are four classes to be taught. In how many different ways can four teachers be chosen for the classes?
13. List the permutations of A, B, C, C, C taking (a) three letters, (b) 4 letters, and (c) 5 letters at a time.
14. Three boys and two girls sit in a row. In how many ways is this possible if the girls (a) must not sit together? (b) must sit at the ends?
15. Five red, three black and two blue counters are put in a row. How many arrangements are possible?
16. How many six-digit numbers can be made with the digits 2, 3, 3, 4, 4, 6? How many of these are odd numbers?
17. Twenty four books are to be placed on a shelf. Twelve of these are paperbacks, four others are volumes of the same set. In how many ways can the shelf be arranged if (a) the set of four volumes are to be together? (b) the paperbacks are to be together and the set of four are to be together? (c) the set of four are to be together in a particular order on the

extreme left and the paperbacks are to be together on the extreme right? (d) the situation is the same as in (c) but the paperbacks are to be in alphabetical order on the extreme right? (Leave answers in factorial form.)

## 3.3 Combinations

1. Express in factorial form

   (a) $\binom{4}{2}$ (b) $\binom{9}{7}$ (c) $\binom{6}{2}$ (d) $\binom{6}{4}$ (e) $\binom{11}{2}$

2. Evaluate (a) $^5C_3$ (b) $^6C_2$ (c) $^9C_5$ (d) $^9C_4$ (e) $^{10}C_3$ (f) $^{50}C_{50}$
3. (a) Show that $^nC_r = {^nC_{n-r}}$ (b) If $^rC_4 = {^rC_2}$ find $r$.
   (c) Show that $^nC_r + {^nC_{r-1}} = {^{n+1}C_r}$ (d) If $^{10}C_r = {^{11}C_{r+1}}$ find $r$.
4. There are twenty different toys in a box. How many selections of three can a boy make?
5. In a school, pupils must choose seven subjects to study. How many possible selections are there if they choose from twenty-five subjects?
6. A team of four is selected from seven players. In how many ways can this be done?
7. There are twelve books on a shelf. How many selections of three books can be made?
8. On a football pool coupon there are thirty matches. In how many ways can ten draws be selected?
9. Five volunteers are required from twelve men. In how many ways can they be chosen?
10. Two teams of seven men are to be formed from fourteen. In how many ways can this be done (a) with no restrictions? (b) if two particular men cannot be in the same team?
11. In a garden centre there are six different shrubs. How many different selections can be made if a man chooses (a) at least one? (b) at least three?
12. A boy must choose at least one book from a shelf containing five books. How many possible selections can he make?
13. If I can choose from seven chocolates and can sample any selection of one or more, how many combinations are possible?
14. In how many ways can a team of three boys and three girls be chosen from six boys and seven girls?
15. I can choose from five cheeses and seven sorts of biscuits. How many selections can I make (a) if I am free to choose any or none of the cheeses and biscuits? (b) if I must choose at least one cheese and at least one sort of biscuit?
16. A bag contains three red balls and two white balls. A second bag contains five blue and four black balls. How many different selections of (a) two balls from the first bag and two from the second, (b) four from the first bag and five from the second, are possible?

## 3.4 Miscellaneous Permutations and Combinations

1. A die is thrown and a coin is tossed. How many different results are possible?
2. A man has three suits, six shirts, four ties, two hats and three pairs of shoes. In how many ways can he be dressed if (a) he must wear one of each item of clothing? (b) he need not necessarily wear a hat or tie, but must wear one of each other item of clothing?
3. In how many ways can a car number plate be made if it is to contain two letters and three numbers if 0 is not allowed as the first number?
4. In how many different ways can a vowel and a consonant be selected from the following words (a) CARPETS? (b) PRETTIER?
5. In a competition there are six questions. Each question has three possible answers. In how many ways can the entry form be completed?
6. (a) In how many ways can a team be formed of a man, a woman, a boy and a girl from two men, six women, three boys and four girls? (b) If a second team is chosen, in how many ways can it be formed?
7. How many numbers have no threes in them between (a) 500 and 599 inclusive? (b) 500 and 650 inclusive? (c) 4000 and 6000 inclusive? (d) How many of the numbers in (a), (b) and (c) are exactly divisible by five?
8. If two numbers are chosen from 1, 1, 2, 3, 3, 5, 5, 8, 9, 10 how many combinations give (a) a sum less than twelve? (b) a sum equal to twelve? (c) a sum greater than twelve?
9. There are three roads from town A to town B each of which has two turnings to town C. There are four roads from town B to town C and two roads direct from A to C. How many routes are there from A to C?
10. In how many ways can six differently coloured beads be put on a bracelet?
11. A committee of five is to be chosen from seven men. How many different arrangements are there if the chosen five (a) sit on a bench? (b) sit round a table?
12. If three letters are to be chosen from the word QUESTION, how many (a) combinations, (b) permutations are possible?
13. Repeat question 12 for the words (a) SERIES (b) QUADRATIC.
14. How many (a) permutations, (b) combinations of four letters of the word PRECIPITATE have exactly two letters the same?
15. Of eighteen cricketers, four are bowlers and five others are good batsmen. How many teams of eleven can be formed choosing precisely two of the bowlers, three of the batsmen and six others?
16. In how many ways can twelve identical sweets be divided among four children?
17. A man can go to work or return by bicycle, car, bus, or on foot. If he goes by car he must return by car. If he goes by bicycle he can return by any means except by car, but his bicycle must be at home on the third night. In how many ways can he go and return for three consecutive days?
18. A boy has 12 pence to spend in a sweet shop. He can choose from three different chocolate bars at 8 p each, four different chocolate bars at 4 p each, six different packets of sweets at 7 p, four at 6 p and two at 5 p. In how many ways can he spend his money if he must spend 20 p exactly?

## 3.5 Probability

**Exercise A**
1. Find the probability of (a) drawing a picture card from a normal pack of 52 cards. (b) obtaining at least one head when three coins are tossed, (c) obtaining an even total score from two dice, (d) obtaining a total score of seven on two dice, (e) obtaining a prime number score on the total of three dice, (f) picking three differently coloured balls together from a bag holding five red, two blue, and two white balls.
2. Draw a tree diagram to show the probabilities and possible outcomes when three balls are taken without replacement from a bag containing six black and nine white balls.
   What are the probabilities that (a) three white balls are taken? (b) two black and one white ball are taken? (c) at least one black ball is taken? (d) an odd number of black balls remains? (e) an even number of white balls remain and at least one black ball is taken? (f) less than five black balls remain and less than eight white balls remain? (g) a white ball is taken on the third trial given that the first two were white?
3. If six married couples are in a room and two people are selected at random find the probability (a) that they are both male, (b) that they are married.
4. A card is selected from a normal pack of 52 cards. Find the probability that it is either a picture card or a heart.
5. A dart board consists of a circle set into a square. The diameter of the circle is of equal length to the side of the square. Find the probability that a dart thrown at random lands in the circle given that it hits the board.
6. Find the probability that an event will happen if the odds are (a) $3:2$, (b) $5:4$, (c) $2:3$ in favour.
7. A die was tossed 200 times. The results were: 25 ones, 28 twos, 35 threes, 24 fours, 54 fives and 34 sixes. Estimate the probability of tossing (a) an even number on one throw, (b) two ones on successive throws, (c) two fives on successive throws.
8. A box contains fifty nails and twenty screws. If two nails are required what is the probability of obtaining at least two if three items are taken at random.
9. Three boys and two girls sit on a seat. Find the probability that (a) the two girls sit together, (b) the two girls sit together and the three boys sit together, (c) the girls are on the ends, (d) the girls are together with boys occupying the end positions.
10. Three letters are chosen at random from the word SEPARATE. Find the probability that (a) three vowels are chosen, (b) two vowels and one consonant are chosen, (c) an A, an E and a T are chosen, (d) an A and an E are chosen in that order, (e) an A is the third letter given that A was the second and T was first, and (f) an A is the third letter given that the first is A.
11. An author writes a good book with a probability of $\frac{1}{2}$. If it is good it is published with a probability of $\frac{2}{3}$. If it is not it is published with a probability of $\frac{1}{4}$. What is the probability that he will get at least one book published if he writes two?
12. If one archer hits a bull's-eye with a probability of $\frac{1}{4}$ and a second hits it

with a probability of $\frac{2}{5}$, what is the probability that (a) the first hits at least one in three shots? (b) the second hits at least two in three shots?

## Exercise B

1. Find the probability that (a) a prime number is thrown on a fair die, (b) if the letters ART are shuffled and put in a row they will form a proper word, (c) the sum of two fair dice is eight when thrown together, (d) if the letters of BAT are arranged at random that B is the middle letter, (e) no heads are thrown when three coins are tossed, (f) two As are chosen from a box of letters containing two As and one B.
2. Draw a tree diagram to illustrate the probabilities of each possible distribution of boys and girls when there are two children in a family, assuming that at each birth a boy or girl is equally likely.
What is the probability (a) of two girls? (b) of two children of the same sex? (c) of a boy and a girl? (d) that the other child is a boy if one is known to be a boy? (e) that the second child is a boy if the first is a boy?
3. If a player is dealt five cards from a normal pack, what is the probability that he will (a) receive no clubs? (b) receive four aces? (c) receive only picture cards?
4. If a registration plate of a car contains three digits what is the probability that two and only two digits are the same?
5. What is the probability that if three dice are thrown (a) three even numbers are obtained? (b) three different numbers are obtained? (c) a total of eight is obtained given that the first die gave an even number?
6. If the letters of HOLLOW are arranged at random what is the probability that (a) they form the word HOLLOW? (b) the two Os come together? (c) the two Os come together given that the Ls are separated?
7. If six people are seated around a table, find the probability that two chosen people are together.
8. If a coin of diameter 1 cm is dropped onto a chess board whose squares are of side 2 cm find the probability that it lies completely in a square.
9. A radio contains five transistors. Each transistor has a probability of $\frac{1}{4}$ of being faulty within six months. If they are not faulty within six months they have a probability of $\frac{1}{10}$ of becoming faulty in the next eighteen months. Find the probability that the radio (a) breaks down within six months, (b) breaks down within two years, (c) will be working after two years, (d) is working after two years given that it survived the first six months.
10. How many times must a fair die be tossed in order to have a probability of at least 0.9 of getting at least one six?
11. A team of four is selected at random from seven players. What is the probability that the best four are selected?
12. Ten raffle tickets are sold for three prizes. A buys three, B buys five, C buys two. Calculate the probability that (a) A wins the first prize, (b) A wins all three prizes, (c) each wins one prize, (d) A wins a prize given that C won the first, (e) A wins a prize given that C wins one prize only.

# 4
# Series

## 4.1 Notes and Formulae

**Finite Series**

In all series work we indicate the series by $u_1, u_2, u_3, \ldots$ and the $r$th term by $u_r$. The sum of $n$ terms is $S_n$ and the sum to infinity is $S_\infty$.

*Arithmetic Progression*

$$u_n = a + (n-1)d$$
$$S_n = \tfrac{1}{2}n\{2a + (n-1)d\}$$
$$= \tfrac{1}{2}n(a+l) \text{ where } l = \text{last term.}$$

Arithmetic mean of $a, b$ is $\tfrac{1}{2}(a+b)$.

*Geometric Progression*

$$u_n = ar^{n-1}$$
$$S_n = \frac{a(r^n - 1)}{r - 1} = \frac{a(1 - r^n)}{1 - r}$$
$$S_\infty = \frac{a}{1-r} \quad \text{valid only if } -1 < r < 1$$

Geometric mean of $a, b$ is $\sqrt{ab}$.

*Binomial Theorem*

$$(a+b)^n = a^n + {}^nC_{n-1} a^{n-1} b + {}^nC_{n-2} a^{n-2} b^2 + \ldots + b^n.$$

(if $n$ is a positive integer).

or $$(1+x)^n = 1 + nx + \frac{n(n-1)}{2!} x^2 + \frac{n(n-1)(n-2)}{3!} x^3 + \ldots + x^n.$$

*Other Series*
Use of the Σ notation. Σ = "sum of all terms of the type ..."

e.g. $\sum_{}^{n} r(r+1) = 1.2 + 2.3 + \ldots + n(n+1)$.

$$\sum_{1}^{n} r = \tfrac{1}{2}n(n+1)$$

$$\sum_{1}^{n} r^2 = \tfrac{1}{6}n(n+1)(2n+1)$$

$$\sum_{1}^{n} r^3 = \tfrac{1}{4}n^2(n+1)^2$$

These results can be combined:

e.g. $\sum r(r+1)^2 = \sum (r^3 + 2r^2 + r)$
$= \sum r^3 + 2\sum r^2 + \sum r$

then use the above results.

$$\sum_{1}^{n} r(r+1) = \tfrac{1}{3}n(n+1)(n+2)$$

$$\sum_{1}^{n} r(r+1)(r+2) = \tfrac{1}{4}n(n+1)(n+2)(n+3)$$

$$\sum \frac{1}{r(r+1)}$$

For series of this type try splitting each term into partial fractions. All but a few terms may then cancel out.

**Infinite Series**

*Convergency*
A series is convergent if $S_n$ tends towards a fixed limit as $n$ tends towards infinity. The geometric series is convergent if $-1 < r < 1$ and the fixed limit (or sum to infinity, written $S_\infty$) is $\dfrac{a}{1-r}$.

A series is convergent if each term can be shown to be numerically less than the corresponding term in a convergent geometric series.

*Geometric Series*
If $S = a + ar + ar^2 + \ldots + ar^{n-1} + \ldots$

$$S_n = \frac{a(r^n - 1)}{r - 1} \quad \text{or} \quad \frac{a(1 - r^n)}{1 - r}$$

If $-1 < r < 1$,

$$S_\infty = \frac{a}{1-r} \quad \text{and} \quad S_\infty - S_n = \frac{ar^n}{1-r}$$

## Binomial Series

$$(1+x)^n = 1 + nx + \frac{n(n-1)}{2!}x^2 + \frac{n(n-1)(n-2)}{3!}x^3 + \cdots$$
$$+ \frac{n(n-1)\cdots(n-r+1)}{r!}x^r + \cdots$$

This series is valid for all values of $n$, positive, negative or fractional, *providing* $-1 < x < 1$.

## Exponential and Logarithmic Series

$$e^x = 1 + x + \frac{x^2}{2!} + \frac{x^3}{3!} + \cdots + \frac{x^r}{r!} + \cdots$$

The exponential series is valid for *all* values of $x$.

$$\log_e(1+x) = x - \tfrac{1}{2}x^2 + \tfrac{1}{3}x^3 - \tfrac{1}{4}x^4 + \cdots + (-1)^{n+1}\frac{x^n}{n} + \cdots$$

The logarithmic series is valid only if $-1 < x \leqslant 1$. Note that there is no series for $\log_e x$ in this range.

$$\log_e(y+1) - \log_e y = 2\left\{\frac{1}{2y+1} + \frac{1}{3(2y+1)^3} + \frac{1}{5(2y+1)^5} + \cdots\right\}$$

This last series is derived from the series for $\log_e(1+x)$ and is valid if $y > 0$.

## Taylor's Theorem
In the region of $x = a$

$$f(x) = f(a) + f_1(a)(x-a) + f_2(a)\frac{(x-a)^2}{2!} + f_3(a)\frac{(x-a)^3}{3!} + \cdots$$
$$+ f_n(a)\frac{(x-a)^n}{n!} + \cdots$$

An alternative form is found by putting $x = a + h$

$$f(a+h) = f(a) + f_1(a)h + f_2(a)\frac{h^2}{2!} + f_3(a)\frac{h^3}{3!} + \cdots + f_n(a)\frac{h^n}{n!} + \cdots$$

## Maclaurin's Theorem (A particular case of Taylor's Theorem.)
In the region of $x = 0$

$$f(x) = f(0) + f_1(0)x + f_2(0)\frac{x^2}{2!} + f_3(0)\frac{x^3}{3!} + \cdots + f_n(0)\frac{x^n}{n!} + \cdots$$

*Sin x and Cos x*

$$\sin x = x - \frac{x^3}{3!} + \frac{x^5}{5!} - \frac{x^7}{7!} + \ldots + (-1)^{n+1}\frac{x^{2n-1}}{(2n-1)!} + \ldots$$

$$\cos x = 1 - \frac{x^2}{2!} + \frac{x^4}{4!} - \frac{x^6}{6!} + \ldots + (-1)^{n+1}\frac{x^{2n}}{(2n)!} + \ldots$$

These are valid for all values of $x$ measured in radians.

The ranges of values of $x$ for which the above results apply depend on the individual series and should be learned by the student.

## 4.2 Arithmetic Progressions

1. Which of the following series are arithmetic? State the common difference for those that are.
   (a) 2, 6, 10, 14 ...
   (b) 4, 2, 0, −2 ...
   (c) 2, 4, 8, 16 ...
   (d) 3, $4\frac{1}{2}$, 6, $7\frac{1}{2}$ ...
   (e) 0, $-1\frac{1}{4}$, $-2\frac{1}{2}$, $-3\frac{3}{4}$ ...
   (f) $\frac{1}{2}$, $\frac{1}{3}$, $\frac{1}{4}$, $\frac{1}{5}$ ...

2. For each arithmetic progression in question 1 find (a) the 10th term, (b) the $n$th term, (c) the sum of 6 terms, (d) the sum of $n$ terms.
3. The second term of an A.P. is 4 and the 6th is 16. Find the first four terms of the progression.
4. In an A.P. the fifth term is 7 and the thirteenth is 17. Find the first term and the common difference.
5. The 5th term of an A.P. is 20 and the 9th is 8. Find the first four terms.
6. The third term of an A.P. is 2, and the fifth is −6. Find the first term and the common difference.
7. The first term of an A.P. is 1 and the sum of 10 terms is 235. Find the common difference.
8. How many consecutive terms of the series 2, 6, 10, ... make a total of 288?
9. The 11th term of an A.P. is −12 and the sum of 11 terms is −22. Find the series.
10. Show that the sum of the first $n$ odd numbers is $n^2$.
11. The first term of an A.P. is −6 and the last term is 30; the sum of all the terms is 156. Find (a) the number of terms, (b) the common difference, and (c) the middle term.
12. An A.P. starts with 5 and the 9th term is 9. Find the number of terms if the total is 210.
13. How many terms of the series 10, 13, 16, ... must be taken for the sum to exceed 400?
14. Find the sum of all the odd numbers from 101 to 199 inclusive.
15. Find the sum of all the even numbers from 52 to 100 inclusive.

## 4.3 Geometric Progressions

1. Which of the following series are geometric? For those that are write down the common ratio.

(a) 8, 12, 18, 27, ...   (b) 48, 36, 27, $20\frac{1}{4}$, ...
(c) 2, $-6$, 18, $-54$, ...   (d) 27, $-9$, $-3$, $-1$, ...
(e) 1, 3, 7, 15, ...   (f) 1, $\frac{1}{3}$, $\frac{1}{9}$, $\frac{1}{27}$ ...

2. In (a) and (c) of the series in question 1 write down the 6th term, the $n$th term, the sum of 6 terms and the sum of $n$ terms.
3. The first term of a G.P. is 5 and the sixth term is 160. Find the sum of 6 terms.
4. The first term of a G.P. is 8 and the 7th term is $\frac{1}{8}$. Find the common ratio and the sum of 7 terms.
5. How many terms are there in the following G.P.s?
   (a) 1, 3, 9, ... 729   (b) 1, $\frac{1}{2}$, $\frac{1}{4}$, ... $\frac{1}{64}$
   (c) $-3$, 6, $-12$ ... 384   (d) 81, 108, 144, ... 256
6. The 3rd term of a G.P. is 36 and the 6th term is $4\frac{1}{2}$. Find the common ratio.
7. How many terms of the series $8 + 24 + 72 + \ldots$ must be taken for the total to be 968?
8. Find the first term of the series 2, 6, 18, ... which exceeds 500.
9. How many terms of the series $2 + 3 + 4\frac{1}{2}$ add up to $\dfrac{665}{16}$?
10. The first 8 terms of a G.P. with a common ratio of 2 add up to 5100. Find the first term.
11. In a G.P. the difference between the 4th term and the 2nd term is $1\frac{1}{2}$, and the difference between the 6th term and the 4th is 4. Find the series.
12. How many terms of the series $81 + 54 + 36 + \ldots$ add up to $221\frac{2}{3}$?
13. The 3rd term of a G.P. is bigger than the first by 45 and the 5th is bigger than the third by 720. Find the series. (Two answers.)

## 4.4 The Binomial Theorem (for Positive Integral Indices)

1. Find the complete expansions of the following:
   (a) $(1 + x)^5$   (b) $(1 + a)^8$   (c) $(1 - z)^7$
2. Find the complete expansions of the following:
   (a) $(x + y)^5$   (b) $(c - d)^6$   (c) $(r + s)^7$
3. Find the coefficients of the terms as far as the middle term (or the first of two middle terms) of the expansions of
   (a) $(x - y)^7$   (b) $(p + q)^9$   (c) $(l - m)^{10}$
4. Find the complete expansions of
   (a) $(1 + 2x)^4$   (b) $(1 - 3y)^6$   (c) $(2 + 3x)^5$   (d) $(4 - x)^4$
5. Find   (a) the 3rd term in the expansion of $(1 + 4x)^9$
   (b) the 5th term in the expansion of $(1 - 3x)^{10}$
   (c) the 6th term in the expansion of $(x + 2y)^7$
   (d) the 5th term in the expansion of $(x - 2y)^{11}$
6. Find the coefficient of $x^3$ in each of the following expansions:
   (a) $(1 + x)^{20}$   (b) $(1 - 2x)^{10}$   (c) $(1 + \frac{1}{2}x)^8$
   (d) $(1 - \frac{1}{3}x)^9$   (e) $(2 - 3x)^7$   (f) $(a - x)^n$
7. Find the coefficient of $x^6$ in the expansions of
   (a) $(1 - x^2)(1 + x)^8$   (b) $(1 + x^3)(1 - 2x)^6$
8. By putting $x + x^2 = y$ expand $(1 + x + x^2)^3$ fully.

9. By putting $2x + x^2 = y$ expand $(1 + 2x + x^2)^5$ as far as the term in $x^3$. Check your result by finding the expansion of $(1 + x)^{10}$ as far as the term in $x^3$.
10. Find the coefficient of $x^3$ in the expansion of $(1 - x)^3 (1 + x + x^2)^4$. (Remember that $1 - x^3 = (1 - x)(1 + x + x^2)$.)
11. Find the coefficient of $x^3$ in the expansion of $(1 + x)^4 (1 - x + x^2)^5$.

## 4.5 Other Series

**Exercise A**

1. Write out in full (a) $\sum_{1}^{5} r^2$ (b) $\sum_{1}^{4} r(r+2)$ (c) $\sum_{1}^{3} \frac{r(r+1)}{r+2}$ (d) $\sum_{1}^{4} (r^3 + 1)$

2. Write down the first four terms of
   (a) $\Sigma 3r^2$ (b) $\Sigma (r+2)(r+4)$ (c) $\Sigma (r^2 + r)$ (d) $\Sigma r(r+1)(r+3)$

3. Write down the following series in $\Sigma$ notation:
   (a) $2 + 8 + 18 + 32 + \ldots$
   (b) $2.3 + 3.4 + 4.5 + \ldots + 10.11$
   (c) $\frac{2.3}{4} + \frac{3.4}{5} + \frac{4.5}{6} + \ldots$
   (d) $3.6.9 + 6.9.12 + 9.12.15 + \ldots + 21.24.27$
   (e) $\frac{1^2}{2} + \frac{2^2}{3} + \frac{3^2}{4} + \ldots + \frac{n^2}{n+1}$

4. Find the sum of the first $n$ terms of the following series:
   (a) $1 + 6 + 15 + \ldots + (2n^2 - n)$
   (b) $4.2 + 5.5 + 6.8 + \ldots + (n+3)(3n-1)$
   (c) $5 + 22 + 63 + \ldots + r(2r^2 + 3) + \ldots$
   (d) $\Sigma 2r^2(2r + 3)$

5. Express $4r^3 + 6r^2 + 6r$ in the form $Ar(r+1)(r+2) + Br(r+1) + Cr$. Using the formulae for $\Sigma r(r+1)(r+2)$, $\Sigma r(r+1)$, and $\Sigma r$ find the sum of the series $\sum_{1}^{n} 4r^3 + 6r^2 + 6r$.

6. Find the sum of the series $\frac{1}{2.3} + \frac{1}{3.4} + \ldots + \frac{1}{(n+1)(n+2)}$.

7. If $S = 1 + 2x + 3x^2 + \ldots + nx^{n-1}$, find $S$ by the following method. Find the series for $xS$ and hence by subtracting one series from the other find the series for $S(1 - x)$; sum the G.P. and hence find $S$.

8. Sum the series $1^2.2 + 2^2.3 + 3^2.4 + \ldots$ to $n$ terms.

**Exercise B**

1. Write down the following series in $\Sigma$ notation:
   (a) $3.1^2 + 5.2^2 + 7.3^2 + \ldots + 15.7^2$
   (b) $\frac{1.2}{3^2} + \frac{3.4}{4^2} + \frac{5.6}{5^2} + \ldots + \frac{19.20}{12^2}$

2. Find the sum of the first $n$ terms of the following series:
   (a) $2 + 6 + 12 + \ldots + (n^2 + n)$
   (b) $3.3 + 4.5 + 5.7 + \ldots + (n+2)(2n+1)$
   (c) $2.1 + 3.3 + 4.5 + \ldots$
   (d) $\Sigma 2r(2r+1)(r+1)$

3. Express $8r^3 + 15r^2 + 9r$ in the form $Ar(r+1)(r+2) + Br(r+1) + Cr$. Using the formulae for summing these expressions find the sum of the series $\sum_{1}^{n} 8r^3 + 15r^2 + 9r$.

4. Show that the sum of the series $\dfrac{1}{2.4} + \dfrac{1}{3.5} + \dfrac{1}{4.6} + \ldots$ to $n$ terms is $\dfrac{1}{2}\left(\dfrac{5}{6} - \dfrac{1}{n+2} - \dfrac{1}{n+3}\right)$.

5. If $S = 1 + x + 2x^2 + 3x^3 + \ldots + (n-1)x^{n-1}$ find the formula for $S$ by the method given in question 7 of the A section.

6. Sum the series $2^2.1 + 3^2.2 + 4^2.3 + \ldots$ to $n$ terms.

## 4.6 Miscellaneous (Finite Series)

1. The sum of the first and third terms of a G.P. is 52, and the sum of the 2nd and 4th terms is 78. Find the first term and the common ratio.

2. A man invests £500 on January 1st of each year and receives compound interest at 10% paid on December 31st of each year. Show that his accumulated capital and interest at the end of $n$ years is $5500(1.1^n - 1)$. Using a calculator (or logarithm tables using $\log_{10} 1.1 = 0.041\ 39$) find the amount of his investment after 20 years to the nearest £10.

3. From a metal rod 95 cm long a series of lengths are cut off. The first piece is 6 cm long and each successive piece is 2 mm shorter than the previous one. Show that 13 pieces can be cut off and find out how much is left over.

4. There are 64 spaces on a chess board. If 1 penny is placed on the first square, 2 p on the second, 4 p on the third and so on doubling each time, estimate the total amount of money on the board when all the spaces are filled.

5. 100 g of a volatile substance is left to evaporate. Each day it loses one tenth of its weight at the beginning of that day. How many days elapse before it is less than half its original weight? Does it follow that it will completely disappear at the end of a similar period of time?

6. Find the sum of 10 terms of the series
   (a) $1.2 + 1.2^2 + 1.2^3 + \ldots$
   (b) $\log_{10} 1.2 + \log_{10} 1.2^2 + \log_{10} 1.2^3 + \ldots$

7. Sum the series
   $1 + (1+x) + (1+x+x^2) + \ldots + (1+x+x^2+\ldots+x^n)$.

8. Write down the expansion of $(a+z)^3$ and then by putting $z = b+c$ find the expansion of $(a+b+c)^3$ and express your answer using the $\Sigma$ notation.

9. Prove, by induction or otherwise, that
$$\sum_{1}^{n} r(r+1)(r+2)(r+3) = \frac{1}{5}n(n+1)(n+2)(n+3)(n+4)$$
10. In a G.P. the product of the first and second terms is one third of the fourth term; the fifth term is the product of the second and third terms. Find the series.
11. An A.P. starts with 4 and the first, second and fifth terms form a G.P. Find the series.

## 4.7 Infinite Geometric Series

1. Find the sum of 6 terms and the sum to infinity of the following:
   (a) $1 + \frac{1}{2} + \frac{1}{4} + \frac{1}{8} + \ldots$
   (b) $2 + \frac{2}{3} + \frac{2}{9} + \frac{2}{27} + \ldots$
   (c) $1 - \frac{1}{2} + \frac{1}{4} - \frac{1}{8} + \ldots$
   (d) $8 + 6 + 4\frac{1}{2} + 3\frac{3}{8} + \ldots$
2. Determine which of the following series are convergent:
   (a) $8 + 4 + 2 + 1 + \ldots$
   (b) $1 + 2 + 4 + 8 + \ldots$
   (c) $25 + 5 + 1 + 0.2 + \ldots$
   (d) $8 + 7 + 6 + 5 + \ldots$
   (e) $1 + x + x^2 + \ldots$ where $-1 < x < 1$.
   (f) $y^4 + y^3 + y^2 + \ldots$ where $y > 1$.
   (g) $4 + 6 + 9 + 13\frac{1}{2} + \ldots$
   (h) $100 - 60 + 36 - 21.6 + \ldots$
   (i) $0.01 + 0.02 + 0.04 + 0.08 + \ldots$
   (j) $1 + 0.1 + 0.01 + 0.001 + \ldots$
   (k) $1 + \frac{1}{y^2} + \frac{1}{y^4} + \ldots$ where $y > 1$.
3. The first term of a geometric series is 100 and the third term is 4. Find its sum to infinity.
4. A geometric series starts with 96 and its sum to infinity is 128. Find the series.
5. The common ratio of a geometric series is $\frac{2}{3}$ and the sum to infinity is 30. Find the first term.
6. How many terms of the series $12 + 6 + 3 + \ldots$ must be taken for the sum to differ from the sum to infinity by less than 0.02.
7. The second term of a geometric series is 20 and the fifth is 1.28. Find the sum to infinity.
8. Find the sum to infinity of the series $1 + \sin^2 \theta + \sin^4 \theta + \ldots$ Evaluate the result when $\theta = 60°$.
9. The second term of a geometric series is 6 and the sum to infinity is 25. Find the first term. (Two answers.)
10. How many terms of the series $108 + 36 + 12 + \ldots$ must be taken for the sum to differ from the sum to infinity by less than 0.05.

11. Find the sum to infinity of the series $1 - \tan^2\theta + \tan^4\theta + \ldots$ where $0 < \theta < 45°$, and evaluate the result if $\theta = 30°$.

## 4.8 The Binomial Series

In the questions in this section assume that the values of $x$ are such that the expansion is valid.

1. Expand the following series as far as the term in $x^3$:
   (a) $(1+x)^{1/2}$  (b) $(1-x)^{-1}$  (c) $(1+x)^{3/4}$  (d) $(1-x)^{-3}$

2. Expand the following expressions as far as the term in $x^3$ and state the permissible range of values for $x$ in each case:
   (a) $(1-2x)^{1/4}$  (b) $(1+3x)^{-3}$  (c) $(1-6x)^{2/3}$  (d) $(1-\frac{1}{2}x)^{-4}$

3. Expand as far as the term in $x^2$ and state the limitation on $x$:
   (a) $\dfrac{1}{\sqrt{1-4x}}$  (b) $(4-3x)^{1/2}$  (c) $(8-5x)^{2/3}$
   (d) $(3-4x)^{-3}$  (e) $\dfrac{1}{\sqrt[3]{1+3x}}$

4. Expand $(1-2x)^{-1/2} - (1-\frac{1}{2}x)^{-2}$ as far as the term in $x^3$.

5. Expand $\sqrt[3]{1+3x} - \dfrac{1}{\sqrt{1+2x}}$ as far as the term in $x^3$. Give the range of permissible values of $x$.

6. Find the first three terms of the following expressions. Give the range of permissible values of $x$.
   (a) $(1+x)\sqrt{1-2x}$  (b) $\dfrac{1-x}{1+x}$  (c) $\sqrt{\dfrac{1+2x}{1-4x}}$  (d) $\dfrac{\sqrt{1-4x}}{(1+x)^2}$

7. Expand as far as the first four terms the expression $\dfrac{1}{(1-x)(1+x)}$ by multiplying out the series for $(1-x)^{-1}$ and $(1+x)^{-1}$. Check your result by comparing it with the expansion of $(1-x^2)^{-1}$.

8. Expand $(1-x)^{-1}$ and hence find an expansion for $(1-x^3)(1-x)^{-1}$. Check your result by another method.

## 4.9 Exponential and Logarithmic Series

1. Write down the first five terms in the expansions of (a) $e^2$ (b) $e^{1/2}$ (c) $e^{-1/3}$

2. Expand as a series (a) $e^{-x}$ (b) $e^{-2x}$ (c) $e^{x/2}$ (d) $e^{-1/x}$ (e) $e^{-x^2}$ (f) $e^{1/x^2}$

3. Write down the value of each of the following series:
   (a) $1 + 5 + \dfrac{25}{2!} + \dfrac{125}{3!} + \ldots$

(b) $1 + \frac{1}{2} + \frac{1}{2!\,2^2} + \frac{1}{3!\,2^3} + \cdots$

(c) $1 + 0.1 + \frac{0.01}{2!} + \frac{0.001}{3!} + \cdots$

(d) $1 - 4 + \frac{16}{2!} - \frac{64}{3!} + \cdots$

(e) $1 - 0.3 + \frac{0.09}{2!} - \frac{0.027}{3!} + \cdots$

4. Express as a series  (a) $\frac{1}{2}(e^3 + e^{-3})$  (b) $\frac{1}{2}(e^4 - e^{-4})$
5. Find the first four terms in the expansion of

(a) $(1+x)e^x$  (b) $(1-2x)e^{-x}$  (c) $\dfrac{e^x + e^{-x}}{2}$

(d) $\dfrac{e^{2x} - 1}{e^x}$  (e) $e^{(x^2+1)}$  (f) $(1+4x)e^{\frac{1}{2}x}$

6. Evaluate the first 14 terms in the expansion of $e^4$. Work to two places of decimals only and devise a system by which each term is calculated from the previous one. Plot the values of the terms on a graph and draw a second graph of the totals as each successive term is added.
7. Expand as a series
   (a) $\log_e(1+\frac{1}{2})$  (b) $\log_e(1.2)$  (c) $\log_e(0.9)$  (d) $\log_e(1+\frac{1}{4})$  (e) $\log_e \frac{7}{8}$
8. Evaluate to 6 places of decimals  (a) $\log_e(1.01)$  (b) $\log_e(0.98)$
9. Multiply together the series for $e^x$ and $e^y$ as far as terms in the third degree and group together terms of a like degree. Show that your expansion is the same as that for $e^{x+y}$.
10. Expand the following as far as the third term, and state in each case the range of values of $x$ for which the series is valid:
    (a) $\log_e(1+2x)$  (b) $\log_e(1-\frac{1}{2}x)$  (c) $\log_e(1+x^2)$
    (d) $\log_e(1-\sqrt{x})$  (e) $\log_e(1-x)^2$  (f) $\log_e(1+2x)(1+x)$
    (g) $\log_e \dfrac{1+2x}{1-x}$  (h) $\log_e(1 - 5x + 6x^2)$

11. If $x$ is so small that $x^4$ and higher powers can be neglected prove that
$$e^x \log_e(1+x) \approx x + \tfrac{1}{2}x^2 + \tfrac{1}{3}x^3$$

12. If $x$ is so small that $x^4$ and higher powers can be neglected prove that
$$e^{-x} \log_e(1+2x) \approx 2x - 4x^2 + \frac{17x^3}{3}$$

## 4.10 Taylor's and Maclaurin's Theorems

1. Write down the first three non-zero terms of  (a) $\sin 2x$  (b) $\cos x^2$
   (c) $\sin \tfrac{1}{3}x$

2. By use of Maclaurin's Theorem verify the expansions of (a) $\sin x$ (b) $\cos x$ (c) $e^x$ (d) $(1+x)^n$ (e) $\log_e(1+x)$
3. Use Maclaurin's Theorem to expand the following series up to the third term:
   (a) $\sin^2 x$   (b) $\tan x$   (c) $e^{\sin x}$
   (d) $e^x \cos x$   (e) $2^x$   (f) $\log_e(1+\sin x)$
4. Find the first four terms in the expansion of
   (a) $(1+x^2)\sin x$   (b) $e^x \sin x$
   (c) $\log_e(1-x)\sin 2x$   (d) $\log_e(1+x+x^2)$
   State the range of validity of these series.
5. Establish which functions give rise to the following series:
   (a) $x^2 - \dfrac{x^4}{3!} + \dfrac{x^6}{5!} - \dfrac{x^8}{7!} + \ldots$

   (b) $1 + x - \dfrac{x^2}{2!} - \dfrac{x^3}{3!} + \dfrac{x^4}{4!} + \dfrac{x^5}{5!} - \dfrac{x^6}{6!} - \dfrac{x^7}{7!} + \ldots$

   (c) $x - \dfrac{x^2}{2!} + \dfrac{x^4}{4!} - \dfrac{x^6}{6!} + \dfrac{x^8}{8!} - \dfrac{x^{10}}{10!} + \ldots$

6. By using the first three non-zero terms of the series for $\sin x$ find approximations for (a) $\sin(0.1 \text{ radians})$ (b) $\sin \tfrac{1}{6}\pi$ (c) $\sin 10°$.
   Check the accuracy of your results against four-figure tables.
7. Show by the use of Taylor's Theorem that in the region of $x = 2$
$$\log_e x = \log_e 2 + \frac{x-2}{2} - \frac{(x-2)^2}{4 \cdot 2!} + \frac{(x-2)^3}{4 \cdot 3!} + \ldots$$
   Find the next two terms.
8. Find the expansion of $\tan^{-1} x$ in the region of $x = 0$ up to the term in $x^7$.
9. Expand $\sin(x+\theta)$ as a power series in $x$.
10. Show that $\tan^{-1}(1+x) = \tfrac{1}{4}\pi + \tfrac{1}{2}x - \tfrac{1}{4}x^2 + \tfrac{1}{12}x^3 + \ldots$

## 4.11 Miscellaneous

1. If $S_n$ is the sum of a geometric series with first term '$a$' and common ratio '$r$', show that
$$\frac{S_{2n} - S_n}{S_n} = r^n.$$

2. Use the Binomial Theorem to expand $\sqrt{1+x}$ and $\dfrac{1}{(1+x)^2}$ as far as the terms in $x^3$. Use these series to obtain values correct to 6 decimal places of $\sqrt{16.08}$ and $\dfrac{1}{(1.005)^2}$.

3. Expand $(1-x)^{1/2}$ as far as the term in $x^3$ and by putting $x = 0.01$ find the value of $\sqrt{11}$ correct to 6 decimal places.

29

4. The first four terms of $(1 + ax)^n$ are $1 - 12x + 90x^2 + px^3 + \ldots$ Find the values of $a$, $n$, and $p$. Deduce the binomial expression producing the series $1 - 6x + 22\frac{1}{2}x^2 - 67\frac{1}{2}x^3 + \ldots$

5. Find the sum to infinity of the series $4 + 3 + 2\frac{1}{4} + \ldots$ If another series has a common ratio of $\frac{7}{8}$ and has the same sum to infinity, find its first term.

6. Show that the difference between the sum of $n$ terms of the series
$$a + ar + ar^2 + \ldots \text{ and the sum to infinity is } \frac{ar^n}{1-r}.$$

7. Expand in series the expression $(1 - x)^{1/4}$ as far as the term in $x^3$. By putting $x = \dfrac{1}{81}$ find the value of $\sqrt[4]{5}$ correct to 5 decimal places.

8. Find the first four terms of the expansion of $\sqrt{1+x}$ and hence evaluate $\sqrt{1.05}$ to 4 decimal places.

The time of swing of a pendulum varies directly as the square root of its length. If the length of a pendulum is increased by 5% find the percentage increase in the time of swing.

9. Find the first two terms in the expansion of $\dfrac{54(1 - \frac{1}{2}x)^{1/4}}{\sqrt{4 - x}(3 + 6x)^3}$

10. Show that $\sqrt{\dfrac{1+x}{1-x}} \approx 1 + x + \frac{1}{2}x^2$. By putting $x = 0.02$ find $\sqrt{51}$ to 4 d.p.

11. Expand the expression $e^{ax} \cdot \dfrac{1 + bx}{1 - 2x}$ as far as the term in $x^3$. If the coefficients of $x$ and $x^2$ are both zero find the values of $a$ and $b$, $(a \neq 0)$. With these values of $a$ and $b$ find the coefficient of $x^3$.

12. Assuming the series for $\log_e(1 + x)$ and by putting $x = \dfrac{1}{y}$ prove that
$$\log_e \frac{y+1}{y-1} = 2\left\{\frac{1}{y} + \frac{1}{3y^2} + \frac{1}{5y^5} + \ldots\right\}$$

For what values of $y$ is the expansion valid? Use the series to find $\log_e 2$ to 4 decimal places.

13. Expand $e^x \log_e(1 + x)$ as a series of ascending powers of $x$ as far as $x^3$ by multiplying out the two series. Can we assume that the next term in the expansion is $\frac{1}{4}x^4$? If not, calculate it. Evaluate the expression $e^{0.1} \log_e(1.1)$ to 6 decimal places and check your answer by using a calculator or tables.

14. Given that $\log_e 10 = 2.3026$, find the value of $\log_{10} 1.1$ to 5 significant figures. (Use five terms of the log series and work to 6 decimal places.)

15. Expand $\sqrt{1-x}$ and $(1 + x)^{-2}$ in the form of two series as far as the term in $x^3$. Hence show that if $x^4$ and higher powers may be neglected then
$$\frac{\sqrt{1-x}}{(1+x)^2} = 1 - 2\frac{1}{2}x + 3\frac{7}{8}x^2 - 5\frac{5}{16}x^3$$

16. Write down the series for $e^x$ and $\sin x^2$. Hence find the first five non-zero terms of $e^x \sin x^2$. By differentiating the result find the first terms in the expansion of $e^x(\sin x^2 + 2x \cos x^2)$.

17. Use Taylor's Theorem to expand $\tan^{-1}(1+x)$ to the first four non-zero terms.
18. If the coefficients of $x$ and $x^2$ are equal in the expansions of $(\sin ax - 3\tfrac{1}{2}x^2)$ and $(e^x + 1)\cos bx$ find $a$ and $b$.

# 5

# Trigonometric Identities and Equations

## 5.1 Notes and Formulae

An **identity** is a relationship true for *all* values of the variable.

An **equation** is a relationship which is true for only *some* values of the variable.

In establishing identities the student starts with one side and transforms it step by step making use of basic identities until the result is the same as the other side. It is acceptable to transform both sides to a common result (working on each separately), or start with a known identity and transform it step by step to the required result.

Geometrical definitions of $\sin\theta$, $\cos\theta$ and $\tan\theta$, for acute angles, are found from a right-angled triangle (with the usual names for the sides).

$$\sin\theta = \frac{\text{OPPOSITE}}{\text{HYPOTENUSE}} \qquad \cos\theta = \frac{\text{ADJACENT}}{\text{HYPOTENUSE}} \qquad \tan\theta = \frac{\text{OPPOSITE}}{\text{ADJACENT}}$$

$\sec\theta$, $\csc\theta$, and $\cot\theta$ are defined as

$$\sec\theta = \frac{1}{\cos\theta} \qquad \csc\theta = \frac{1}{\sin\theta} \qquad \cot\theta = \frac{1}{\tan\theta}$$

Use of Pythagoras' Theorem establishes

$$\sin^2\theta + \cos^2\theta = 1 \qquad \sec^2\theta = 1 + \tan^2\theta \qquad \csc^2\theta = 1 + \cot^2\theta$$

These identities are true for *any* angle.

To establish the trigonometrical ratios for angles bigger than 90° we imagine the plane split into four quadrants by $x$ and $y$ axes. A rotating arm swings from the positive $x$-axis through the required angle, $\theta$ (positive is anticlockwise). If the arm is of length '$r$' (taken always as positive) and its end point is $(x, y)$ then

$$\sin\theta = \frac{y}{r} \qquad \cos\theta = \frac{x}{r} \qquad \tan\theta = \frac{y}{x}$$

taking the sign of $x$ and $y$ into account.

In all cases a short cut is to find the acute angle made between the arm and the x-axis. The numerical value of the required ratio is the same as for this angle. The sign to be taken is according to the diagram which gives the ratios which are positive in the four quadrants. It is always wise, however, to have a mental picture of the direction of the arm and visualise whether the end of the arm is up or down and to the left or the right.

| Sin | All |
|---|---|
| Tan | Cos |

Angles can be in degrees or radians, where one radian is the angle subtended by an arc of a circle equal to the radius.

$$\pi \text{ radians} = 180°$$

If the trig. ratio of a *number* is referred to it should be interpreted as that number of radians.

$$\text{e.g. } \sin \pi = \sin (\pi \text{ radians}) = 0.$$

Standard results:

$$\sin 0 = 0 \qquad \cos 0 = 1 \qquad \tan 0 = 0$$

$$\sin 30° = \frac{1}{2} \qquad \cos 30° = \frac{\sqrt{3}}{2} \qquad \tan 30° = \frac{1}{\sqrt{3}}$$

$$\sin 45° = \frac{1}{\sqrt{2}} \qquad \cos 45° = \frac{1}{\sqrt{2}} \qquad \tan 45° = 1$$

$$\sin 60° = \frac{\sqrt{3}}{2} \qquad \cos 60° = \frac{1}{2} \qquad \tan 60° = \sqrt{3}$$

$$\sin 90° = 1 \qquad \cos 90° = 0 \qquad \tan 90° = \infty$$

The student should be able to deduce that

$$\cos(90° - x) = \sin x, \qquad \tan(90° - x) = \cot x, \qquad \sin(90° - x) = \cos x.$$

and results in all quadrants such as

$$\sin(90° + x) = \cos x, \qquad \sin(180° + x) = -\sin x, \qquad \cos(180° - x) = -\cos x.$$

## 5.2 Graphs and Ratios for any Angle

**Exercise A**
1. Find in surd form
   (a) $\sin 225°$    (b) $\cos(-270°)$    (c) $\tan 135°$    (d) $\sin 120°$
   (e) $\cos 330°$    (f) $\sec 210°$    (g) $\tan(-60°)$    (h) $\cot 765°$
   (i) $\sin(-300°)$    (j) $\operatorname{cosec} 315°$
2. Find all angles between 0 and $2\pi$ which satisfy
   (a) $\sin \theta = \frac{\sqrt{3}}{2}$    (b) $\cos \theta = -\frac{1}{2}$    (c) $\tan \theta = -1$

(d) $\sec \theta = \sqrt{2}$ (e) $\operatorname{cosec} \theta = -2$ (f) $\cot \theta = \sqrt{3}$
(g) $\sin \theta = -\dfrac{1}{\sqrt{2}}$ (h) $\cos \theta = \dfrac{1}{\sqrt{2}}$

3. Find using tables
   (a) $\cos 120°$ (b) $\cos 330°$ (c) $\tan 275°$ (d) $\cot 117°$
   (e) $\cos(-190°)$ (f) $\operatorname{cosec} 342°$ (g) $\sin(-230°)$ (h) $\sec(-217°)$
   (i) $\cos 297°$ (j) $\sin 700°$ (k) $\cos(-920°)$ (l) $\tan 627°$

4. Convert into radians
   (a) $35°$ (b) $112°$ (c) $193°$ (d) $355°$ (e) $712°$ (f) $840°$

5. Convert into degrees and minutes
   (a) $\dfrac{2\pi}{3}$ radians (b) $\dfrac{5\pi}{8}$ radians (c) $\dfrac{5\pi}{3}$ radians (d) 1 radian
   (e) 1.2 radians (f) 1.5 radians (g) 0.75 radians (h) 3.5 radians

6. Sketch graphs of
   (a) $y = \sin x$ (b) $y = \cos x$, for $-720° \leq x \leq 720°$.

7. Sketch (a) $y = \operatorname{cosec} \theta$ (b) $y = \sec \theta$, for $-360° \leq \theta \leq 360°$.

8. Find (a) $\sin \dfrac{3\pi}{4}$ (b) $\cos\left(-\dfrac{5\pi}{2}\right)$ (c) $\tan \dfrac{4\pi}{5}$
   (d) $\cot\left(-\dfrac{7\pi}{5}\right)$ (e) $\operatorname{cosec} \dfrac{3\pi}{8}$ (f) $\sec 1.25\pi$
   (g) $\tan 0.75$ (h) $\sin(-2)$ (i) $\cos 3.4$
   (j) $\operatorname{cosec}(-2, 6)$

9. Find all the angles between $-360°$ and $+360°$ which satisfy
   (a) $\sin x = 0.5$ (b) $\cos x = -0.2306$ (c) $\tan x = 3.7583$
   (d) $\operatorname{cosec} x = -2\tfrac{1}{2}$ (e) $\sec x = \tfrac{4}{3}$ (f) $\sin x = 0.3812$
   (g) $\cot x = -0.2731$ (h) $\cos x = 0.5731$ (i) $\operatorname{cosec} x = -1.718$
   (j) $\sin x = 0.4444$

10. Find all the angles between $-180°$ and $+180°$ which satisfy
    (a) $\cos(x + 30°) = 0.631$ (b) $\sin(x - 50°) = 0.2781$
    (c) $\cos(180° - x) = -0.728$ (d) $\sin(360° + x) = 0.8872$
    (e) $\tan(180° + 2x) = 3.681$ (f) $2 \cot x = 3$
    (g) $\operatorname{cosec}^2 x = 2.74$ (h) $\cos 2x = -0.487$
    (i) $\sin^2 3x = 0.7123$ (j) $\cot^2(x + 30°) = 2.87$
    (k) $3 \cos^2(2x + 50°) = 2$

11. Find all the values of $\theta$ from $0°$ to $360°$ which satisfy
    (a) $\sin \theta (\sin \theta - 1) = 0$ (b) $(2 \cos \theta + 1)(3 \cos \theta - 2) = 0$
    (c) $2 \cos^2 \theta + 3 \cos \theta = 0$ (d) $6 \sin^2 \theta - 7 \sin \theta - 3 = 0$
    (e) $\tan^2 \theta + \tan \theta = 2$ (f) $\sec^2 \theta = 5 \sec \theta$
    (g) $\operatorname{cosec} \theta = \sin \theta$ (h) $2 \sin \theta = 3 \cos \theta$
    (i) $3 \tan \theta = 4 \sin \theta$ (j) $3 \sin \theta - 2 \operatorname{cosec} \theta = 5$
    (k) $2 \cos^2 \theta = 7 \sin^2 \theta$ (l) $8 \sin^2 \theta - 2 \sin \theta - 1 = 0$
    (m) $4 \sin^2 \theta - \tan^2 \theta = 0$ (n) $\dfrac{5 \sin \theta}{1 + 2 \sin \theta} = 1$
    (o) $\dfrac{3 \cos \theta}{1 + \cos \theta} = 1 - \cos \theta$

33

12. Simplify
    (a) $\sin(90° + \theta)$
    (b) $\cos(180° + \theta)$
    (c) $\tan(360° - \theta)$
    (d) $\cot(180° - \theta)$
    (e) $\tan(720° + \theta)$
    (f) $\cot(90° + \theta)$
    (g) $\sin(270° - \theta)$
    (h) $\tan(270° + \theta)$
    (i) $\sec(270° + \theta)$
    (j) $\cosec(-\theta)$
    (k) $\tan(540° - \theta)$
    (l) $\tan(-\theta)$

**Exercise B**
1. Sketch (a) $y = \tan x$ (b) $y = \cot x$ for $-2\pi \leqslant x \leqslant 2\pi$
2. Sketch (a) $y = \sin x + \cos x$ (b) $y = x + \sin x$ for $0 \leqslant x \leqslant 4\pi$
3. Find all the angles between 0 and $2\pi$ which satisfy
    (a) $\cos\theta = \dfrac{1}{\sqrt{2}}$
    (b) $\sin\theta = -\dfrac{\sqrt{3}}{2}$
    (c) $\tan\theta = -\dfrac{1}{\sqrt{3}}$
    (d) $\sec\theta = -\sqrt{2}$
    (e) $\cosec\theta = \sqrt{2}$
    (f) $\tan\theta = \sqrt{3}$
    (g) $\sin\theta = -\dfrac{1}{\sqrt{2}}$

4. Find all the angles between $-2\pi$ and $+2\pi$ satisfying
    (a) $\cos x = 0.5$
    (b) $\sin x = 0.3146$
    (c) $\tan x = -1.87$
    (d) $\cosec x = \dfrac{7}{3}$
    (e) $\sec x = \dfrac{7}{5}$
    (f) $\sin x = 0.8791$
    (g) $\cot x = -0.3415$
    (h) $\cos x = 0.2794$
    (i) $\cosec x = -2.3842$
    (j) $\sin x = 0.6677$

5. Find all angles between $-180°$ and $180°$ which satisfy
    (a) $\cos(x + 60°) = 0.7523$
    (b) $\sin(x - 20°) = 0.3651$
    (c) $\tan(180° - x) = -3.5$
    (d) $\sin(360° + x) = 0.5789$
    (e) $\cot(180° - 2x) = 1.892$
    (f) $2\cos 3x = 0.78$
    (g) $\cosec^2 2x = 3.587$
    (h) $\cot^2(x - 30°) = 5.921$
    (i) $4\sin^2(3x - 50°) = 1$
    (j) $\cos 5x = -0.1768$
    (k) $4\sin^2(160° - 3x) = 3$

6. Find values of $\theta$ between 0 and 360° which satisfy
    (a) $\cos\theta(1 + \cos\theta) = 0$
    (b) $(2\sin\theta - 1)(\sin\theta + 1) = 0$
    (c) $3\tan^2\theta + 2\tan\theta = 0$
    (d) $2\cos^2\theta - 3\cos\theta + 1 = 0$
    (e) $\tan^2\theta + 3\tan\theta - 4 = 0$
    (f) $3\sin\theta = \cos\theta$
    (g) $2\cot\theta = 3\cos\theta$
    (h) $5\cot\theta = \tan\theta$
    (i) $2\sec\theta + \cos\theta - 3 = 0$
    (j) $2\cos^2\theta - 3\sin^2\theta = 0$
    (k) $5\sin^2\theta + 2\sin\theta - 2 = 0$
    (l) $2\cos^2\theta - 5\cot^2\theta = 0$
    (m) $\dfrac{3\sin\theta}{2 - 4\sin\theta} = 1$
    (n) $\dfrac{2\cos\theta + 1}{1 - 2\cos\theta} = 1 + 2\cos\theta$

7. Simplify
    (a) $\cos(90° - \theta)$
    (b) $\sin(180° + \theta)$
    (c) $\tan(270° - \theta)$
    (d) $\cot(180° + \theta)$
    (e) $\sec(720° + \theta)$
    (f) $\cos(90° + \theta)$
    (g) $\sec(-\theta)$
    (h) $\dfrac{\sin(180° + \theta)}{\cos(360° - \theta)}$
    (i) $\cot(-\theta)$
    (j) $\dfrac{\tan(180° + \theta)}{\cot(360° - \theta)}$

## 5.3 Identities and Use of Basic Formulae

**Exercise A**

1. Show that
   (a) $\dfrac{\sin\theta}{\sqrt{1-\sin^2\theta}} = \tan\theta$
   (b) $\dfrac{\cos\theta}{1-\sin^2\theta} = \sec\theta$
   (c) $1+\tan^2\theta = \sec^2\theta$
   (d) $1+\cot^2\theta = \operatorname{cosec}^2\theta$
   (e) $\dfrac{\tan\theta}{\sqrt{1+\tan^2\theta}} = \sin\theta\cos\theta$
   (f) $(\operatorname{cosec}^2\theta - 1)(1+\tan^2\theta) = \operatorname{cosec}^2\theta$
   (g) $a^2\cos^2\theta + b^2\sin^2\theta = a^2 + (b^2 - a^2)\sin^2\theta$
   (h) $\tan\theta + \cot\theta = \sec\theta\operatorname{cosec}\theta$
   (i) $\sin^4\theta + 2\sin^2\theta\cos^2\theta + \cos^4\theta = 1$

2. Prove the following identities:
   (a) $\tan\theta\sec\theta = \sin\theta(1+\tan^2\theta)$
   (b) $(\sec\theta - 1)(\sec\theta + 1)\cot\theta = \tan\theta$
   (c) $\sec^4\theta - \tan^4\theta = \tan^2\theta + \sec^2\theta$
   (d) $\cos^4\theta - \sin^2\theta = \sin^4\theta - \cos^2\theta$
   (e) $\dfrac{\sqrt{1+\cot^2\theta}}{\sin\theta} = \dfrac{\sqrt{1+\tan^2\theta}}{\cos\theta}\cot^2\theta$
   (f) $\dfrac{1}{1-\cos\theta} + \dfrac{1}{1+\cos\theta} = 2\operatorname{cosec}^2\theta$
   (g) $\dfrac{2\tan\theta}{1+\tan^2\theta} = 2\sin\theta\cos\theta$
   (h) $\dfrac{1-\tan^2\theta}{1+\tan^2\theta} = 2\cos^2\theta - 1$
   (i) $\dfrac{\sec\theta + \operatorname{cosec}\theta}{\tan\theta + \cot\theta} = \dfrac{\tan\theta - \cot\theta}{\sec\theta - \operatorname{cosec}\theta}$

3. Eliminate $\theta$ from
   (a) $x = a\cos\theta,\ y = a\sin\theta$
   (b) $x = b\tan\theta,\ y = a\sec\theta$
   (c) $x = \tan\theta,\ y = a\sin\theta$
   (d) $x = \sin\theta + \cos\theta,\ y = \sin\theta - \cos\theta$
   (e) $x+y = \cot\theta,\ x-y = \tan\theta$
   (f) $x = \sec\theta\tan\theta,\ y = \cot\theta$
   (g) $x = a + \tan\theta,\ y = \sec\theta$

4. Find the smallest positive values of $\theta$ which make the following functions a maximum and a minimum, and find the corresponding values of the function.

   (a) $1 + 3\sin\theta$
   (b) $4 - 2\cos\theta$
   (c) $\dfrac{1}{\cos\theta + 1}$
   (d) $\cos^2 2\theta$
   (e) $\dfrac{1}{5 - 3\sin 2\theta}$
   (f) $\dfrac{4}{\sec(\theta - 30°)}$

5. If $\theta$ is acute, find without using tables the value of
   (a) $\tan\theta$ if $\sin\theta = \dfrac{3}{5}$
   (b) $\cos\theta$ if $\tan\theta = \dfrac{5}{12}$

35

(c) $\sin \theta$ if $\sec \theta = \dfrac{17}{8}$  (d) $\cosec \theta$ if $\tan \theta = \dfrac{4}{3}$

**6.** Find a value of $\theta$ between $0°$ and $360°$ satisfying

(a) $\sin \theta = 0$, and $\cos \theta = -1$  (b) $\tan \theta = -1$ and $\sin \theta = -\dfrac{1}{\sqrt{2}}$

(c) $\cot \theta = \dfrac{1}{\sqrt{3}}$ and $\sec \theta = -2$  (d) $\cosec \theta = \dfrac{-2}{\sqrt{3}}$ and $\sec \theta = -2$

### Exercise B

**1.** Show

(a) $(\cos \theta + \sin \theta)^2 + (\cos \theta - \sin \theta)^2 = 2$

(b) $\dfrac{\cot^2 \theta}{\cosec \theta - 1} = \cosec \theta + 1$

(c) $\cos^2 \theta \dfrac{1 - \cos \theta}{1 - \sin \theta} = \sin^2 \theta \dfrac{1 + \sin \theta}{1 + \cos \theta}$

(d) $\sin^2 \theta = \dfrac{\sec \theta - \cos \theta}{\sec \theta}$

(e) $\dfrac{\sin \theta}{\tan \theta - 1} + \dfrac{\cos \theta}{\cot \theta - 1} = 0$

(f) $\dfrac{1}{\cos^2 \theta} - \dfrac{1}{\sin^2 \theta} = (\tan \theta - \cot \theta)(\tan \theta + \cot \theta)$

**2.** Prove the following identities:

(a) $\dfrac{\cosec A - \cot A}{\sec A - \tan A} = \dfrac{\sec A + \tan A}{\cosec A + \cot A}$

(b) $\dfrac{1}{1 + \tan A - \sec A} = \dfrac{(1 + \tan A + \sec A)}{2 \tan A}$

(c) $\dfrac{1}{\cot A + \tan A} = \sin A \cos A$

(d) $\dfrac{\sin A + \cos A}{1 + 2 \sin A \cos A} = \dfrac{1}{\sin A + \cos A}$

(e) $\dfrac{1 + \sin A}{1 - \sin A} = (\sec A + \tan A)^2$

(f) $\dfrac{1}{1 + \tan^2 A} = 1 - \dfrac{1}{1 + \cot^2 A}$

(g) $\cot^4 A - \tan^4 A + \sec^4 A - \cosec^4 A = 2(\tan^2 A - \cot^2 A)$

(h) $(\sin A + \cos A)^3 - (\sin A - \cos A)^3 = 2 \cos A (3 - 2 \cos^2 A)$

(i) $\dfrac{1}{(\sec A - \tan A)(\cosec A - \cot A)} = 1 + \dfrac{1 + \sin A + \cos A}{\sin A \cos A}$

**3.** Eliminate $\theta$ from

(a) $x = 3 \sin \theta$, $y = 4 \cos \theta$

(b) $x = 2\sin\theta + 3\cos\theta$, $y = 3\sin\theta - 2\cos\theta$
(c) $x = 2\sec\theta$, $y = \cot\theta$
(d) $x = \sec\theta - \tan\theta$, $y = \sec\theta + \tan\theta$
(e) $x + y = \tan\theta$, $y = \cos\theta$ \quad (f) $x\sin\theta = y$, $y\cos\theta = x + y$

4. Find the (i) maximum, (ii) minimum value and the smallest positive value of $\theta$ to give these values for the functions:

(a) $2 - \sin\theta$ \quad (b) $3\cos\theta - 1$ \quad (c) $\dfrac{2}{3 + 2\cos 2\theta}$

(d) $3\sin^2\tfrac{1}{4}\theta$ \quad (e) $\dfrac{1}{4 - 2\cos^2\tfrac{3}{2}\theta}$

5. If $\theta$ is acute find
(a) $\sin\theta$ if $\cos\theta = \tfrac{1}{4}$ \quad (b) $\tan\theta$, $\sec\theta$ if $\sin\theta = \tfrac{2}{3}$
(c) $\cot\theta$, $\csc\theta$ if $\cos\theta = \tfrac{1}{5}$ \quad (d) $\sin\theta$, $\sec\theta$ if $\tan\theta = \tfrac{2}{7}$

6. Find the values of $\theta$ between $-180°$ and $+180°$ which satisfy

(a) $\sin\theta = \dfrac{\sqrt{3}}{2}$ and $\cos\theta = -\dfrac{1}{2}$ \quad (b) $\tan\theta = 1$ and $\cos\theta = -\dfrac{1}{\sqrt{2}}$

(c) $\sec\theta = 2$ and $\sin\theta = -\dfrac{\sqrt{3}}{2}$ \quad (d) $\cot\theta = -\sqrt{3}$ and $\sin\theta = \dfrac{1}{2}$

# 6

# Trigonometric Formulae

## 6.1 Notes and Formulae

**Addition Formulae**

$$\sin(A + B) = \sin A \cos B + \cos A \sin B$$
$$\sin(A - B) = \sin A \cos B - \cos A \sin B$$
$$\cos(A + B) = \cos A \cos B - \sin A \sin B$$
$$\cos(A - B) = \cos A \cos B + \sin A \sin B$$
$$\tan(A + B) = \frac{\tan A + \tan B}{1 - \tan A \tan B}$$
$$\tan(A - B) = \frac{\tan A - \tan B}{1 + \tan A \tan B}$$

$$\cos 2A = \cos^2 A - \sin^2 A = 2\cos^2 A - 1 = 1 - 2\sin^2 A$$
$$\sin 2A = 2\sin A \cos A$$
$$\sin 3A = 3\sin A - 4\sin^3 A \qquad \cos 3A = 4\cos^3 A - 3\cos A$$

$$\tan 2A = \frac{2\tan A}{1-\tan^2 A} \qquad \tan 3A = \frac{3\tan A - \tan^3 A}{1-3\tan^2 A}$$

$$\sin 2A = \frac{2\tan A}{1+\tan^2 A} \qquad \cos 2A = \frac{1-\tan^2 A}{1+\tan^2 A}$$

**Sums and Differences**

$$\sin A + \sin B = 2\sin\tfrac{1}{2}(A+B)\cos\tfrac{1}{2}(A-B)$$
$$\sin A - \sin B = 2\sin\tfrac{1}{2}(A-B)\cos\tfrac{1}{2}(A+B)$$
$$\cos A + \cos B = 2\cos\tfrac{1}{2}(A-B)\cos\tfrac{1}{2}(A+B)$$
$$\cos A - \cos B = -2\sin\tfrac{1}{2}(A-B)\sin\tfrac{1}{2}(A+B)$$

conversely
$$2\sin P \cos Q = \sin(P+Q) + \sin(P-Q)$$
$$2\cos P \cos Q = \cos(P+Q) + \cos(P-Q)$$
$$2\sin P \sin Q = \cos(P-Q) - \cos(P+Q)$$

**Small Angles**

If $\theta$ is so small that $\theta^3$ can be neglected then

$$\sin\theta \approx \theta \quad \text{and} \quad \cos\theta \approx 1 - \tfrac{1}{2}\theta^2$$

**Solution of $a\cos\theta + b\sin\theta = c$**

*Method 1*

Put $a\cos\theta + b\sin\theta \equiv R\cos(\theta - \alpha)$ where $R = \sqrt{a^2+b^2}$

$$\cos\alpha = \frac{a}{\sqrt{a^2+b^2}} \quad \text{and} \quad \sin\alpha = \frac{b}{\sqrt{a^2+b^2}}$$

or put $a\cos\theta + b\sin\theta \equiv R\sin(\theta + \alpha)$ where $R = \sqrt{a^2+b^2}$

$$\sin\alpha = \frac{a}{\sqrt{a^2+b^2}} \quad \text{and} \quad \cos\alpha = \frac{b}{\sqrt{a^2+b^2}}$$

*Method 2*

Put $t = \tan\tfrac{1}{2}\theta$, so that

$$\sin\theta = \frac{2t}{1+t^2}; \quad \cos\theta = \frac{1-t^2}{1+t^2}. \quad \left(\text{Also } \tan\theta = \frac{2t}{1-t^2}.\right)$$

Substitute these values and solve the quadratic in $t$.

**Inverse Functions**

If $\sin^{-1} x = A$, then $A = \sin x$, with similar definitions for the other trigonometrical ratios.

To solve identities with inverse functions use the normal identities and invert at the appropriate point.

**General Solutions**

If $\alpha$ is one solution of
    (a) $\sin x = a$,    then    $x = n\pi + (-1)^n \alpha$,
    (b) $\cos x = a$,    then    $x = 2n\pi \pm \alpha$

(c) $\tan x = a$, then $x = n\pi + \alpha$, are the general solutions, $n$ being any integer.

To solve equations use the standard formulae to express the equation in factors each containing only one ratio. Caution: do not cancel a ratio from an equation or one of the solutions will be lost.

e.g. If $2 \sin x \cos x = \cos x$ do *not* cancel $\cos x$ from both sides write as

$$\cos x (2 \sin x - 1) = 0$$

giving $\cos x = 0$, or $\sin x = \frac{1}{2}$

whence $x = 2n\pi \pm \frac{\pi}{2}$ or $x = n\pi + (-1)^n \frac{\pi}{6}$

## 6.2 Addition Formulae

### Exercise A

1. Find without the use of tables or calculator
   (a) $\sin 15°$  (b) $\cos 165°$  (c) $\tan 285°$
   (d) $\sec 105°$  (e) $\sin 75°$  (f) $\tan 105°$
2. Simplify
   (a) $\sin(x + 30°)$  (b) $\cos(45° - x)$  (c) $\cos(90° + x)$
   (d) $\tan(270° - x)$  (e) $\cot(180° + x)$  (f) $\sin(330° - x)$
3. Find without the use of tables or calculator
   (a) $\frac{1}{2} \cos 15° + \frac{\sqrt{3}}{2} \sin 15°$  (b) $\frac{1}{\sqrt{2}}(\cos 15° - \sin 15°)$
4. From the formulae for $\sin(A + B)$ and $\cos(A + B)$ show that
   (a) $\tan(A + B) = \dfrac{\tan A + \tan B}{1 - \tan A \tan B}$  (b) $\cos 2A = \cos^2 A - \sin^2 A$
   (c) $\sin 2A = 2 \sin A \cos A$
5. If $A$ and $B$ are acute and $\tan A = \dfrac{5}{12}$, $\tan B = \dfrac{3}{4}$
   Find without the use of tables or calculator
   (a) $\tan(A + B)$  (b) $\cos(A - B)$  (c) $\cot 2A$
6. If $\cos A = \dfrac{3}{5}$ and $\cos B = \dfrac{5}{13}$, and $A$ and $B$ are acute find without the use of tables or calculator
   (a) $\sin(A + B)$  (b) $\cos(A - B)$  (c) $\tan(A + B)$
7. Simplify
   (a) $\sin 25° \cos 16° - \cos 25° \sin 16°$  (b) $\dfrac{\tan 27° + \tan 18°}{1 - \tan 27° \tan 18°}$
   (c) $\dfrac{1 - \tan 15°}{1 + \tan 15°}$  (d) $\dfrac{\sqrt{3}}{2} \cos x - \dfrac{1}{2} \sin x$
   (e) $\dfrac{1}{\sqrt{2}} \cos x + \dfrac{1}{\sqrt{2}} \sin x$  (f) $2 \sin 15° \cos 15°$

(g) $\cos^2 30° - \sin^2 30°$      (h) $\cos^2 30° + \sin^2 30°$

(i) $\dfrac{1 - \tan^2 15°}{2 \tan 15°}$

**8.** Find all the values of $x$ between $0°$ and $180°$ which satisfy
   (a) $\cos x = 2 \cos(x + 60°)$      (b) $\sin(x + 45°) = 2 \cos(x - 30°)$
   (c) $\sin(x - 30°) = \frac{3}{4} \sin(x - 40°)$      (d) $2 \sin 2x = 3 \cos^2 x$
   (e) $\sin 3x = 2 \cos(3x - 10°)$

**9.** Show that $\sin(A - 30°) + \cos(A + 60°) = 0$.

**10.** If $\sin(x + 30°) = \cos(x - 45°)$ find $\tan x$.

**11.** If $\tan A = 2$ and $\tan(A - B) = \frac{1}{4}$ find $\tan B$.

**12.** If $\cos \theta = \dfrac{5}{13}$ find $\sin 2\theta$ and $\cos 2\theta$, $\theta$ being acute.

**13.** If $\tan(A + 30°) = 2$ find $\tan A$.

**14.** Simplify    (a) $2 \cos^2 22\frac{1}{2}° - 1$    (b) $\cos^2 15° - \sin^2 15°$

**15.** Prove the following identities:
   (a) $\sin(x + y) - \sin(x - y) \equiv 2 \cos x \sin y$
   (b) $\cos(x + y) + \cos(x - y) \equiv 2 \cos x \cos y$
   (c) $\dfrac{\sin(A + B)}{\cos A \cos B} \equiv \tan A + \tan B$
   (d) $\dfrac{\cos(A - B)}{\cos A \cos B} \equiv 1 + \tan A \tan B$
   (e) $\dfrac{\cos 2\theta}{\cos \theta - \sin \theta} \equiv \cos \theta + \sin \theta$
   (f) $(\cos \theta + \sin \theta)^2 - (\cos \theta - \sin \theta)^2 \equiv 2 \sin 2\theta$
   (g) $\cot A - \tan A \equiv 2 \cot 2A$
   (h) $\cot B + \tan B \equiv \cot B \sec^2 B$
   (i) $\dfrac{\cos(A + B)}{\sin A \cos B} + \dfrac{\sin(A - B)}{\sin A \sin B} \equiv 2 \cot 2B$
   (j) $\tan(x - y) + \tan y \equiv \dfrac{\sin x}{\cos y \cos(x - y)}$

**16.** Solve, for $0 \leqslant x \leqslant 360°$:
   (a) $\sin(x + 30°) = \cos(x - 45°)$      (b) $2 \cos(x + 45°) = \cos x$
   (c) $\tan(x - 25°) = \tan 15°$

**17.** Show that
$$\tan(A + B + C) = \dfrac{\tan A + \tan B + \tan C - \tan A \tan B \tan C}{1 - \tan A \tan B - \tan B \tan C - \tan C \tan A}.$$

**18.** Prove the following identities:
   (a) $\tan A - \tan B \equiv \dfrac{\sin(A - B)}{\cos A \cos B}$
   (b) $\tan A + \cot A \equiv 2 \operatorname{cosec} 2A$
   (c) $\cos(A + B) + \cos(A - B) \equiv 2 \cos A \cos B$
   (d) $\dfrac{\cos(A - B) - \cos(A + B)}{\sin(A + B) + \sin(A - B)} \equiv \tan B$

(e) $\sin 2\theta + \cos 2\theta + 1 \equiv 2\cos\theta(\sin\theta + \cos\theta)$
(f) $2\cos^2 A \cot 2A \equiv \cot A \cos 2A$
(g) $\sec 2A - \tan 2A \equiv \dfrac{1-\tan A}{1+\tan A}$
(h) $\operatorname{cosec} 2A - \cot 2A \equiv \tan A$

19. If $A, B, C$ are angles of a triangle show that
   (a) $\tan A + \tan B + \tan C = \tan A \tan B \tan C$
   (b) $\sin A + \sin B + \sin C = 4\cos\tfrac{1}{2}A \cos\tfrac{1}{2}B \cos\tfrac{1}{2}C$

20. Show that
   (a) $\dfrac{\sin 2\theta}{2\cos\theta} = \sin\theta$
   (b) $\dfrac{\cos 2\theta}{\cos\theta - 1} = \cos\theta + 1$
   (c) $\cot 2\theta = \tfrac{1}{2}(\cot\theta - \tan\theta)$
   (d) $\dfrac{2\cos^2\theta}{\cos 2\theta + 1} = 1$
   (e) $\sin 2\theta = \dfrac{2\tan\theta}{1+\tan^2\theta}$
   (f) $\cot 2\theta = \dfrac{\cot^2\theta - 1}{2\cot\theta}$
   (g) $\sin\theta + \sin 3\theta = 4\sin\theta \cos^2\theta$ (h) $\cos\theta - \cos 3\theta = 2\cos\theta \sin^2\theta$
   (i) $\cos\theta + \cos 3\theta = 2\cos\theta \cos 2\theta$
   (j) $(\sqrt{2}\cos 2\theta - 1)(\sqrt{2}\cos 2\theta + 1) = \cos 4\theta$
   (k) $1 + \dfrac{1-\cos 4\theta}{1+\cos 4\theta} = \sec^2 2\theta$

**Exercise B**

1. Find in surd form  (a) $\sin 75°$  (b) $\tan 15°$  (c) $\cos 345°$
   (d) $\cot 255°$  (e) $\cos 15°$  (f) $\cot(-15°)$  (g) $\cos 7\tfrac{1}{2}°$

2. If $A$ and $B$ are acute and $\sin A = \dfrac{8}{17}$ and $\cos B = \dfrac{3}{5}$ find
   (a) $\cos(A-B)$  (b) $\sin(A+B)$  (c) $\tan 2A$

3. Find $\tan(A+45°) - \tan(A-45°)$ in terms of $\tan A$.

4. Simplify
   (a) $\dfrac{\sqrt{3}}{2}\sin x + \dfrac{1}{2}\cos x$  (b) $\sin x - \sqrt{3}\cos x$
   (c) $\dfrac{1+\tan x}{1-\tan x}$  (d) $\sin 12° \cos 18° + \cos 12° \sin 18°$
   (e) $\dfrac{1+\sqrt{3}\tan x}{\sqrt{3}-\tan x}$

5. Find $\tan x$ if $\tan(A+x) = 2$, and $\tan A = \tfrac{1}{4}$.
6. Find $\tan x$ if $\sin(x+\theta) = \cos(x-\theta)$.
7. Simplify
   (a) $\dfrac{1}{2}\cos x - \dfrac{\sqrt{3}}{2}\sin x$  (b) $\cos x + \sin x$
   (c) $\dfrac{\sqrt{3}-\tan x}{1+\sqrt{3}\tan x}$  (d) $2\cos^2 15° - 1$

(e) $\dfrac{1}{2}\sin\dfrac{1}{2}x\cos\dfrac{1}{2}x$  (f) $\dfrac{2\tan^2 22\frac{1}{2}°}{1-\tan^2 22\frac{1}{2}°}$

8. If $\sin\theta = \dfrac{7}{25}$ find $\sin 2\theta$, $\cos 3\theta$, $\tan 2\theta$, and $\sin 3\theta$.
9. Show $\sin\theta - \sin(\theta - 60°) = \sin(\theta + 60°)$.
10. Use the formulae for $\sin(A+B)$, $\cos(A+B)$ and $\tan(A+B)$ to show
    (a) $\cos 2A = \cos^2 A - \sin^2 A = 1 - 2\sin^2 A = 2\cos^2 A - 1$
    (b) $\sin 3A = 3\sin A - 4\sin^3 A$  (c) $\cos 3A = 4\cos^3 A - 3\cos A$
    (d) $\tan 3A = \dfrac{3\tan A - \tan^3 A}{1 - 3\tan^2 A}$

11. Solve
    (a) $\sin(x - 60°) = \sin x$  (b) $\cos(x + 45°) = 2\sin(x - 45°)$
    (c) $2\tan(x + 30°) + \tan(x + 120°) = 0$, for $-180° \leqslant x \leqslant 180°$.
12. Show that
    (a) $\cos 2A = \dfrac{1 - \tan^2 A}{1 + \tan^2 A}$  (b) $\sin 2A = \dfrac{2\tan A}{1 + \tan^2 A}$
    (c) $\tan A = \dfrac{2\tan\frac{1}{2}A}{1 - \tan^2\frac{1}{2}A}$  (d) $\sin A = \dfrac{2\tan\frac{1}{2}A}{1 + \tan^2\frac{1}{2}A}$
13. Show that $\dfrac{\sin A}{1 + \cos A} = \dfrac{1 - \cos A}{\sin A}$.
14. If $A + B + C = 90°$, find $\tan(A + B + C)$ and hence show that $\tan A \tan B + \tan B \tan C + \tan C \tan A = 1$.
15. Prove the following identities:
    (a) $\cos B \sin A + \sin(B - A) = \sin(A + B) - \sin A \cos B$
    (b) $\dfrac{\sin A}{\cos B} + \dfrac{\cos A}{\sin B} = \dfrac{2\cos(A - B)}{\sin 2B}$
    (c) $\dfrac{\sin A}{\cos B} + \dfrac{\sin B}{\cos A} = \dfrac{\sin 2A + \sin 2B}{\cos(B + A) + \cos(B - A)}$
    (d) $\dfrac{\cos A + \sin A}{\cos A - \sin A} = \tan(45° + A)$
    (e) $\sec(A + B) = \dfrac{\sec A \sec B}{1 - \tan A \tan B}$
    (f) $\sec 2A = 1 + \tan A \tan 2A$
    (g) $\dfrac{\cos A + \sin A}{\cos A - \sin A} - \dfrac{\cos A - \sin A}{\cos A + \sin A} = 2\tan 2A$
    (h) $\sec^2 A(1 + \sec 2A) = 2\sec 2A$
    (i) $\cos 4A = 1 - 8\cos^2 A + 8\cos^4 A$
    (j) $\tan 3\theta \tan 2\theta \tan\theta = \tan 3\theta - \tan 2\theta - \tan\theta$
    (k) $\dfrac{\sin\theta + \sin 2\theta}{1 + \cos\theta + \cos 2\theta} = \tan\theta$
    (l) $\tan 2\theta = \tan\theta(\sec 2\theta + 1)$
    (m) $\cos^4\theta + \sin^4\theta = \frac{1}{2}(1 + \cos^2 2\theta)$

(n) $\csc A - 2\sin A = 2\cot 2A \cos A$
16. (a) If $\sin x = 0.2$, find $\cos 2x$ and $\sin 3x$.
    (b) If $\cos 2x = \frac{1}{4}$, find $\sin 2x$, $\sin x$, and $\cos x$.
    (c) Find $\sin 22\frac{1}{2}°$ and $\cos 22\frac{1}{2}°$ in surd form.
17. Solve for $0 \leq \theta \leq 360°$
    (a) $\dfrac{\sin 2\theta}{1 + \cos 2\theta} = \sqrt{3}$
    (b) $\csc 2\theta + \cot 2\theta = \frac{1}{2}$
    (c) $\dfrac{1 + \sin \theta - \cos \theta}{1 + \sin \theta + \cos \theta} = 2$
    (d) $\csc \theta - 2 \cot 2\theta \cos \theta = 1$
    (e) $\dfrac{\cos x - \sin x}{\cos x + \sin x} = 3$
    (f) $\sin \theta + \sin 3\theta = \cos \theta + \cos 3\theta$
    (g) $\csc \theta + \cot \theta = \tan 2\theta$
    (h) $\tan 2\theta + 2 \cot 2\theta = 3 \csc 2\theta$
18. (a) If $\tan x = \frac{1}{2}$, find $\tan 2x$.
    (b) If $\cos 2x = \frac{1}{3}$, find $\sin x$ and $\cos x$.
    (c) Find $\sin 2\theta$ and $\cos 2\theta$ if $\sin \theta = \dfrac{5}{13}$ and $\theta$ is obtuse.
    (d) Find $\tan 22\frac{1}{2}°$ in surd form.
19. Solve for $0 \leq \theta \leq 360°$
    (a) $\sin \theta = \cos 2\theta + 1$
    (b) $3 \sin 2\theta = 2 \sin \theta$
    (c) $\sin 3\theta + \cos 2\theta - 1 = 0$
    (d) $6 \tan \theta = \tan 2\theta$
    (e) $3 \tan \theta = 2 \tan 3\theta$

## 6.3 Sums and Differences

**Exercise A**
1. Write as products
   (a) $\cos 60° + \cos 30°$
   (b) $\sin 40° - \sin 20°$
   (c) $\cos 40° - \sin 20°$
   (d) $\sin(x + 30°) - \sin x$
   (e) $\cos(x - h) - \cos x$
   (f) $\sin(x + h) + \sin(x - h)$
2. Write as sums or differences of trigonometrical ratios
   (a) $2 \sin 60° \cos 30°$
   (b) $\sin 3x \sin 7x$
   (c) $2 \cos(A - B) \cos(A + B)$
   (d) $2 \sin(x + y - z) \cos(y + z - x)$
   (e) $\cos(3\pi + \frac{1}{2}x) \cos(3\pi - \frac{1}{2}x)$
3. Use the formulae for $\sin(A \pm B)$, $\cos(A \pm B)$ to show that
   (a) $\sin P + \sin Q = 2 \sin \frac{1}{2}(P + Q) \cos \frac{1}{2}(P - Q)$
   (b) $\cos P - \cos Q = 2 \sin \frac{1}{2}(P + Q) \sin \frac{1}{2}(Q - P)$
4. Show that
   (a) $\sin 40° - \sin 80° + \sin 20° = 0$
   (b) $16 \sin 20° \sin 40° \sin 60° \sin 80° = 3$
5. Show that
   (a) $\sin A + \sin 2A + \sin 3A = \sin 2A (2 \cos A + 1)$
   (b) $\dfrac{\cos A + \cos B}{\sin A - \sin B} = \cot \frac{1}{2}(A - B)$
   (c) $\dfrac{\sin A + \sin B}{\sin A - \sin B} = \dfrac{\tan \frac{1}{2}(A + B)}{\tan \frac{1}{2}(A - B)}$

(d) $1 + 2\cos(135° - x)\cos(45° + x) = \sin 2x$
(e) $\frac{1}{2}\sec(45° + x)\sec(45° - x) = \sec 2x$

6. Prove the following identities:
   (a) $\dfrac{\sin 5A - \sin 3A}{\cos 5A + \cos 3A} \equiv \tan A$

   (b) $\dfrac{\cos 2B - \cos 2A}{\sin 2B + \sin 2A} \equiv \tan(A - B)$

   (c) $\dfrac{\sin A + \sin B}{\cos A + \cos B} \equiv \tan\tfrac{1}{2}(A + B)$

   (d) $\cos A \cos 2A \cos 3A \equiv \dfrac{1}{8}\left(1 + \dfrac{\sin 7A}{\sin A}\right)$

   (e) $\sin A + 2\sin 3A + \sin 5A \equiv 2\sin 3A(1 + \cos 2A)$
   (f) $\sin A + 2\sin 3A + \sin 5A \equiv 2\cos A(\sin 2A + \sin 4A)$
   (g) $\cos\dfrac{\pi}{8}\cos\dfrac{3\pi}{8} - \cos\dfrac{5\pi}{8}\cos\dfrac{7\pi}{8} = 0$
   (h) $\cos x \cos\tfrac{1}{2}x - \cos 2x \cos\tfrac{3}{2}x = \sin x \sin 5x$
   (i) $(\cos 6x - \cos 2x)\cos 2x + (\sin 6x + \sin 2x)\sin 2x = 0$

7. If $A$, $B$, $C$, are angles of a triangle show that
   (a) $\cos A + \cos(B - C) = 2\sin C \sin B$
   (b) $\cos\tfrac{1}{2}(A - C) - \sin\tfrac{1}{2}B = 2\sin\tfrac{1}{2}A \sin\tfrac{1}{2}C$
   (c) $\cos A + \cos B + \cos C = 1 + 4\sin\tfrac{1}{2}A \sin\tfrac{1}{2}B \sin\tfrac{1}{2}C$
   (d) $\cos 2A + \cos 2B + \cos 2C = -(1 + 4\cos A \cos B \cos C)$
   (e) $\tan A + \tan B + \tan C = \tan A \tan B \tan C$

8. Solve the following equations for $0° \leqslant x \leqslant 360°$;
   (a) $\sin x + \sin 2x = 0$  (b) $\cos 3x - \cos 4x = 0$
   (c) $\sin(x + 30°) - \sin x = 0$
   (d) $\cos(3x + 20°) + \cos(3x - 40°) = 0$
   (e) $\sin(x + 10°) + \sin(x - 10°) = 1$
   (f) $\cos(x - 20°) - \cos(x + 15°) = 0.2$
   (g) $\sin 3x + \cos x = 0$
   (h) $2\sin(x + 20°)\cos(x - 20°) = 0.3$
   (i) $2\cos(2x + 15°)\cos(2x - 20°) = 0.1$
   (j) $\sin(3x - 12°)\sin(3x + 15°) = 0.8$
   (k) $\sin x + \sin 3x + \sin 5x = 0$

## Exercise B

1. Write as products
   (a) $\sin 2x + \sin 4x$   (b) $\cos 4x + \cos 8x$
   (c) $\sin 7x - \sin 5x$   (d) $\sin\dfrac{5}{2}x + \sin\dfrac{1}{2}x$
   (e) $\dfrac{\sqrt{3}}{2} + \sin x$   (f) $\sin x + \cos x$
   (g) $\sin 90° + \sin 2x$   (h) $\sin(x - 60°) + \sin(x + 60°)$
   (i) $\cos(x + 15°) - \cos(x - 75°)$

2. Write as sums or differences of trigonometrical ratios
   (a) $2 \cos 40° \cos 20°$
   (b) $\cos(3x + \alpha) \sin(3x - \alpha)$
   (c) $\sin 2x \sin 4x$
   (d) $2 \sin(\frac{1}{2}x + 15°) \cos(\frac{1}{2}x - 15°)$
   (e) $2 \cos \frac{3\pi}{8} \cos \frac{\pi}{8}$
   (f) $\cos \frac{5}{3}y \sin \frac{4}{3}y$

3. Use the formulae for $\sin A \pm \sin B$, and $\cos A \pm \cos B$ to show that
   (a) $2 \sin P \cos Q = \sin(P+Q) + \sin(P-Q)$
   (b) $2 \cos P \cos Q = \cos(P+Q) + \cos(P-Q)$
   (c) $\sin P \sin Q = \frac{1}{2}\cos(P-Q) - \frac{1}{2}\cos(P+Q)$

4. Show that (a) $\cos 40° + \cos 80° - \cos 20° = 0$
   (b) $16 \cos 20° \cos 40° \cos 60° \cos 80° = 1$

5. Show that
   (a) $\cos A + \cos 2A + \cos 3A = \cos 2A(2 \cos A + 1)$
   (b) $\dfrac{\sin A + \sin B}{\cos A + \cos B} = \tan \frac{1}{2}(A+B)$
   (c) $\dfrac{\sin 3A - \sin A}{\cos 3A + \cos A} = \tan A$
   (d) $4 \cos A \cos(A + 60°) \cos(A + 120°) = -\cos 3A$
   (e) $\frac{1}{2} \cosec(45° + A) \cosec(45° - A) = \sec 2A$

6. Prove the following identities:
   (a) $\dfrac{\sin A - \sin B}{\cos A + \cos B} \equiv \tan \frac{1}{2}(A-B)$
   (b) $\dfrac{\sin A + \sin 2A}{\cos A - \cos 2A} \equiv \cot \frac{1}{2}A$
   (c) $\dfrac{\sin 7A + \sin 5A}{\cos 5A + \cos 7A} \equiv \tan 6A$
   (d) $\dfrac{\sin x + 2 \sin 3x + \sin 5x}{\cos x + 2 \cos 3x + \cos 5x} \equiv \tan 3x$
   (e) $\sin x + \cos x + \sin 3x + \cos 3x \equiv 2 \cos x (\sin 2x + \cos 2x)$
   (f) $\dfrac{\sin(A + 2B) + \sin A}{\cos(A + 2B) - \cos A} \equiv -\cot B$
   (g) $\dfrac{\cos(x - 30°) + \sin x}{\cos(60° + x) + \cos x} \equiv \tan(30° + x)$
   (h) $2 \sin(45° + x) \sin(45° - x) \equiv \cos 2x$
   (i) $\sin(A - B) \sin(A + B) \equiv \sin A \sin(A + 2B) - \sin B \sin(B + 2A)$
   (j) $\cos \frac{1}{2}\theta \sin \frac{7}{2}\theta - \sin \frac{3}{2}\theta \cos \frac{5}{2}\theta \equiv \sin 2\theta \cos \theta$

7. If A, B, C, are the angles of a triangle show that
   (a) $\sin B - \sin(A - C) = 2 \cos A \sin C$
   (b) $\sin \frac{1}{2}A + \cos \frac{1}{2}(B - C) = 2 \cos \frac{1}{2}B \cos \frac{1}{2}C$
   (c) $\sin A + \sin B + \sin C = 4 \cos \frac{1}{2}A \cos \frac{1}{2}B \cos \frac{1}{2}C$
   (d) $\sin 2A + \sin 2B + \sin 2C = 4 \sin A \sin B \sin C$
   (e) $\cos 2A + \cos 2B - \cos 2C = 1 - 4 \sin A \sin B \cos C$

8. Solve the following equations for $-180° \leqslant x \leqslant 180°$:
   (a) $\cos(x - 12°) + \cos(x + 12°) = 1.2$
   (b) $\sin(x - 2°) + \sin(x + 10°) = 0.8$
   (c) $\cos 2x + \sin x = 0$
   (d) $2 \sin(x - 30°) \cos(x + 30°) = 0.25$

(e) $2\sin(3x+16°)\sin(3x-12°) = 0.4$
(f) $\sin(2x-12°)\sin(2x+80°) = 0.15$
(g) $\cos x + \cos 3x + \cos 5x = 0$
(h) $\sin 2x + \sin 4x + \sin 6x = 0$
(i) $\cos x + 2\cos 3x + \cos 5x = 0$
(j) $\sin 2x + \sin 4x + \sin 3x + \sin 5x = 0$
(k) $\sin \frac{2}{5}x + 2\sin \frac{4}{5}x + \sin \frac{9}{5}x = 0$

## 6.4 Small Angles and Graphical Solutions

1. Using the fact that $\sin \theta \approx \theta$ for small angles in radians show that
   (a) $\cos \theta \approx 1 - \frac{1}{2}\theta^2$; (b) $\tan \theta \approx \theta$ (c) $\sec \theta \approx 1$
   (d) $2 - 2\cos \theta \approx \theta^2$ (e) $\frac{1}{2}(1 - \cos 2\theta) \approx \theta^2$

2. Solve graphically for $0 \leqslant \theta \leqslant \frac{1}{2}\pi$:
   (a) $\cos \theta = \theta$ (b) $\tan 2\theta = 6\theta - 1$
   (c) $\cos \theta - 2\sin \theta = -1$ (d) $2\cos^2 \theta = \theta + 1$

3. If $\theta$ is small and in radians show that
   (a) $\cos 2\theta \approx 1 - 2\theta^2$ (b) $\dfrac{\sin 4\theta}{3\theta} \approx \dfrac{4}{3}$
   (c) $\dfrac{\sin \theta \sin \frac{1}{2}\theta}{1 - \cos \theta} \approx 1$ (d) $\dfrac{\theta \sin \theta}{1 - \cos 3\theta} \approx \dfrac{2}{9}$
   (e) $\dfrac{\sin 5\theta}{\sin 6\theta} \approx \dfrac{5}{6}$

4. If $\theta$ is small and in radians find approximations for
   (a) $\sin 2\theta$ (b) $\cos 3\theta$ (c) $\dfrac{\sin 5\theta}{1 - \cos 4\theta}$
   (d) $\dfrac{\theta^2}{1 - \cos 2\theta}$ (e) $\dfrac{\theta \sin 2\theta}{1 - \cos \frac{1}{2}\theta}$ (f) $\cos(\theta + \frac{1}{2}\pi)$
   (g) $\sin^2 \theta \cot \theta$ (h) $\dfrac{\tan(\theta + h) - \tan h}{\theta}$ (i) $\sec^2 \theta \cos 2\theta$

5. If $\theta$ is small and in *degrees* show that
   (a) $\sin \theta \approx \dfrac{\pi \theta}{180}$ (b) $\cos \theta \approx 1 - \left(\dfrac{\pi}{180}\right)^2 \dfrac{\theta^2}{2}$

6. Solve the following equations graphically:
   (a) $\theta + \frac{1}{4}\pi = \cos \theta$ for $-\frac{1}{2}\pi \leqslant \theta \leqslant \frac{1}{2}\pi$
   (b) $\sin 2\theta - \frac{1}{2}\theta = 0$ for $0 \leqslant \theta \leqslant \frac{1}{2}\pi$
   (c) $\cos(2\theta - \frac{1}{3}\pi) = 2\theta - 1$ for $0 \leqslant \theta \leqslant \pi$
   (d) $\sec \theta = 2 - \frac{1}{4}\theta$ for $0 \leqslant \theta \leqslant \frac{1}{2}\pi$

7. If $\theta$ is small and in radians find approximations for
   (a) $\dfrac{\cos(\theta + h) - \cos(\theta - h)}{\theta}$ (b) $\dfrac{1 - \cos 3\theta}{1 - \cos 4\theta}$
   (c) $\dfrac{\sin \theta}{\theta}$ (d) $1 - \dfrac{\theta \sin 8\theta}{\cos 4\theta}$

(e) $\dfrac{\cos\theta}{1-2\sin^2\frac{1}{2}\theta}$   (f) $\dfrac{\sin^2 2\theta - 1}{2\theta - 1}$

## 6.5 Miscellaneous

1. Draw graphs of  (a) $\tan^{-1} x$  (b) $\sin^{-1} x$  (c) $\cos^{-1} 2x$  (d) $2\sin^{-1} 3x$ for one period of the function.
2. If $\sin^{-1} A = 60°$ and $\sin^{-1} B = 20°$ find
   (a) $\cos^{-1} A$   (b) $\sin^{-1}(A-B)$   (c) $\cos(\cos^{-1} A + \cos^{-1} B)$
3. If $\cos^{-1} A = 30°$ and $\sin^{-1} B = 15°$ find
   (a) $\sin^{-1}\sqrt{1-A^2}$   (b) $\tan^{-1} A$   (c) $\tan^{-1}\dfrac{B}{A}$
   (d) $\sin^{-1}(AB)$   (e) $\tan(2\sin^{-1} A - \cos^{-1} B)$
4. Show that $\tan^{-1} x + \tan^{-1} y = \tan^{-1}\dfrac{x+y}{1+xy}$.
   Use this to show that   (a) $\tan^{-1} 3 + \tan^{-1} 2 = \tfrac{3}{4}\pi$
   (b) $\tan^{-1}\tfrac{1}{2} + \tan^{-1}\tfrac{1}{4} + \tan^{-1}\tfrac{7}{6} = \tfrac{1}{2}\pi$
5. If $\tan^{-1} x + \tan^{-1} y + \tan^{-1} z = \pi$, show that $x + y + z = xyz$.
6. Find $x$ if $\tan^{-1}\tfrac{1}{2} - \tan^{-1}\tfrac{1}{3} = \cos^{-1} x$.
7. Show that $\cot^{-1} A + \cot^{-1} B = \cot^{-1}\dfrac{AB-1}{A+B}$. Hence show that
$$\cot^{-1}(\sqrt{2}+1) - \cot^{-1}(\sqrt{2}-1) = \operatorname{cosec}^{-1}\sqrt{2}.$$
8. Write in the form $R\cos(\theta + \alpha)$
   (a) $3\cos\theta + 4\sin\theta$   (b) $2\cos\theta - 3\sin\theta$
9. Write in the form $R\sin(2\theta + \alpha)$
   (a) $5\sin 2\theta + 12\cos 2\theta$   (b) $6\sin 2\theta - 3\cos 2\theta$   (c) $8\cos 2\theta - 3\sin 2\theta$
10. Find the maximum and minimum values of the following and the minimum positive value of $\theta$ at which they occur.
    (a) $3\cos\theta + 4\sin\theta$   (b) $\sqrt{3}\sin\theta + \cos\theta$   (c) $(\cos\theta + 3\sin\theta)^2$
    (d) $\cos 2\theta - \sin 2\theta$   (e) $-(5\sin 3\theta + 12\cos 3\theta)$
11. By writing in the form $R\cos(\theta + \alpha)$ or $R\sin(\theta + \alpha)$ solve for $0° \leqslant \theta \leqslant 360°$
    (a) $3\cos\theta - 4\sin\theta = 5$   (b) $2\cos\theta + 3\sin\theta = 1$
    (c) $5\sin\theta + 2\cos\theta = 3$   (d) $\sin\theta = \tfrac{1}{4} - \cos\theta$
    (e) $3\cos\theta = 4\cos\theta - 5\sin\theta + 1$   (f) $2\tan\theta = 3\sec\theta - 2$
12. Substitute $t = \tan\theta$ to find solutions for $0° \leqslant \theta \leqslant 360°$ of
    (a) $\cos 2\theta + 2\sin 2\theta = 1$   (b) $5\sin 2\theta - 3\cos 2\theta = 2$
    (c) $5\cos 2\theta - 12\sin 2\theta = 3$   (d) $4\sin 2\theta - 3\cos 2\theta = 2$
13. Substitute $t = \tan\tfrac{1}{2}\theta$ to solve the following equations for $-180° \leqslant \theta \leqslant 180°$
    (a) $\cos\theta + 2\sin\theta = 1$   (b) $3\sin\theta - 4\cos\theta = 3$
    (c) $5\sin\theta + 7\cos\theta = 4$   (d) $3\cos\theta - 6\sin\theta = 2$

14. Find general solutions to
    (a) $\sin 2\theta = \sin 3\theta$
    (b) $\sin(2\theta + \frac{1}{3}\pi) = \cos(2\theta + \frac{1}{3}\pi)$
    (c) $2\cos 2\theta + 2\cos\theta = 1$
    (d) $2\sin 3\theta = \sin\theta$
    (e) $3\sin\theta + 4\cos\theta = 2$
    (f) $\sin 5\theta + \sin\theta = \sin 3\theta$
    (g) $\cos 4\theta + \cos 2\theta = \cos 3\theta$
    (h) $\sin(2\theta + 60°) + \sin(3\theta - 20°) = 0$
    (i) $\sin\theta + \sin 2\theta + \sin 3\theta + \sin 4\theta = 0$
    (j) $\sin 2\theta = \cos\theta \sin 3\theta$
    (k) $\tan(2\theta + \frac{1}{6}\pi) + \tan(4\theta + \frac{1}{6}\pi) = 0$

# 7

# Solution of Triangles

## 7.1 Notes and Formulae

**Notation**
Angles are described by capital letters $A$, $B$, $C$. Sides opposite the angles are given by the corresponding lower case letters, $a$, $b$, $c$.

**Sine Rule**

$$\frac{a}{\sin A} = \frac{b}{\sin B} = \frac{c}{\sin C} = 2R$$

where $R$ is the radius of the circumcircle.

**Cosine Rule**

$$\cos A = \frac{b^2 + c^2 - a^2}{2bc}$$

and by transformation
$$a^2 = b^2 + c^2 - 2bc\cos A$$
similarly
$$b^2 = c^2 + a^2 - 2ca\cos B$$
and
$$c^2 = a^2 + b^2 - 2ab\cos C.$$

**Area of Triangle ABC**
Area of triangle ABC $= \frac{1}{2}ab\sin C$ (or $\frac{1}{2}bc\sin A$, or $\frac{1}{2}ca\sin B$)
$$= \sqrt{s(s-a)(s-b)(s-c)} \text{ where } s = \frac{1}{2}(a+b+c).$$

If the sides are put into order of magnitude, the corresponding angles are in the same order of magnitude.

The area of a sector of a circle or radius '$r$' and angle $\theta$ radians is $\frac{1}{2}\theta r^2$. The length of the arc of the sector is $\theta r$.

48

When solving triangles use the appropriate formulae to give the most accurate answers. The cosine rule involves more calculations than the sine rule, so choose the latter if possible.

*Particular Cases*
When two angles are known, subtract from 180° for the third.
(a) Three sides given. Find one angle (preferably the largest) by the cosine rule and the second by the sine rule.
(b) Two sides and included angle given. Find the third side by the cosine rule and another angle by the sine rule.
(c) Two sides and an angle opposite one of them (this is the *ambiguous case*). Use the sine rule to find a second angle (there may be one, two or no possible solutions); subtract the two angles from 180°; find the third side by the sine rule.
(d) One side and two angles given. Obtain the third angle by the use of the sine rule.
(e) Two or three angles given. Insoluble, but the sine rule gives the *ratios* between the sides.

**Alternative Formulae**
Use of the following formulae is an alternative to the cosine rule when three sides are known:

$$\sin \tfrac{1}{2} A = \sqrt{\frac{(s-b)(s-c)}{bc}} \qquad \cos \tfrac{1}{2} A = \sqrt{\frac{s(s-a)}{bc}}$$

$$\sin A = \frac{2}{bc} \sqrt{s(s-a)(s-b)(s-c)} \qquad \tan \tfrac{1}{2} A = \sqrt{\frac{(s-b)(s-c)}{s(s-a)}}$$

The relationship $c = a \cos B + b \cos A$ is often useful in solving problems on triangles.

**Three Dimensions**
A line is perpendicular to a plane if it is perpendicular to every line in the plane.
The angle between a line and a plane is the angle between the line and its projection in the plane.
The angle between two intersecting planes is the angle between two lines drawn perpendicular to the line of contact in the two planes. It is the angle of greatest slope and is equal to the angle between the normals to the planes.
In solving problems in three dimensions, first draw a clear diagram and then reduce the problem to a succession of two dimensional studies in the planes of the solid.

## 7.2 Solution of Triangles and Miscellaneous Areas

1. Obtain the sine rule $\dfrac{a}{\sin A} = \dfrac{b}{\sin B} = \dfrac{c}{\sin C}$.
2. Obtain the cosine rule $\cos A = \dfrac{a^2 + b^2 - c^2}{2bc}$.

3. Solve the following triangles (i.e. find all the remaining sides and angles) and find their areas:
   (a) $A = 54°, B = 38°, a = 12$ cm.
   (b) $B = 125°, A = 12°, a = 15$ m.
   (c) $b = 12.2$ mm, $c = 7.4$ mm, $B = 46° 12'$.
   (d) $a = 16$ cm, $B = 27° 17', C = 112° 11'$.
   (e) $B = 12° 20', C = 102° 10', a = 17.5$ cm.
   (f) $a = 11$ cm, $b = 14$ cm, $c = 20$ cm.
   (g) $a = 6.2$ m, $b = 7.1$ m, $c = 5.4$ m.
   (h) $b = 12$ m, $c = 11$ m, $C = 18° 9'$.
   (i) $a = 14.7$ cm, $A = 67° 24', B = 40° 30'$.
4. An arc $AB$ subtends an angle of $40°$ at the centre $O$ of a circle of radius 3 cm. Find the area between the straight line $AB$ and the arc $AB$, and also the perimeter of this shape.
5. Repeat question 4 where the angle is $\theta°$ and the radius is '$r$' cm.
6. Find the area cut from a circle radius '$a$' between two parallel chords, one of length '$a$' and the other of length $\frac{1}{2}a$ on the same side of the centre.
7. Find the area between parallel chords of a circle of radius '$a$' which subtend angles of $40°$ and $70°$ respectively at the centre and which are on the same side of the centre.
8. Find the area between arc $AB$ and chord $AB$ of a circle of radius '$a$' when the chord $AB = \frac{1}{5}a$ in length. Find also the length of arc $AB$.
9. If a chord $AB$ subtends an angle $\theta$ at the centre of a circle of radius '$a$', show that the triangle $OAB$ has an area $\frac{1}{2}a^2 \sin\theta$.
10. Find the area between a hexagon of side length '$a$' and its circumscribing circle.
11. A square $ABCD$ whose sides are of length 3 units is drawn. The arc of a circle whose centre, $O$, is the centre of the square and whose radius is $OA$ is drawn from $A$ to $B$. Find the area of the whole figure.
12. An isosceles triangle $ABC$ is drawn. $A = 50°, B = C$, and $AB = 6$ cm. Arc $BC$ is cut off from a circle of radius 3 cm. Find the area of the figure bounded by $AC$, $BA$, and arc $BC$.
13. Chord $AB$ of a circle of radius '$a$' is of length $\frac{1}{2}a$. Find the ratio of the areas of the major sector $ABO$ and minor sector $ABO$, where $O$ is the centre of the circle.
14. A chord $AB$ divides a circle into two parts whose areas are in the ratio $3:1$. If $AB$ subtends an angle $\theta$ radians at the centre show that $\theta - \sin\theta = \frac{1}{2}\pi$.
15. Two circles have their centres 12 cm apart. They have radii 6 cm and 8 cm. Find the size of the area common to the two circles.
16. A chord $AB$ subtends an angle $\theta$ radians at the centre of a circle. A parallel chord $CD$ between $AB$ and the centre subtends an angle of $2\theta$ radians. Find the ratio of the area between $AB$ and $CD$ to that of the rest of the circle.

## 7.3 Miscellaneous Problems with Triangles

**Exercise A**
1. In triangle $ABC$, $D$ is the mid-point of $BA$. Angles $BCD = \alpha$, and $DCA = \beta$. Show that $b(\sin\beta - \sin A) = a(\sin\alpha - \sin B)$.

2. In triangle $ABC$, $D$ is the point on $AB$ such that $DC$ bisects angle $C$. If $BD = p$ and $DA = q$, show by use of the sine rule that

$$\frac{p}{q} = \frac{a}{b}$$

3. In triangle $ABC$, $CD$ is the perpendicular from $C$ to $AB$. If $AD = l$, and $DB = m$, show that $(a^2 + b^2) - (l^2 + m^2) = 2ab \sin A \sin B$.
4. In triangle $ABC$ show that
   (a) $a = b \cos C + c \cos B$   (b) $a(b \cos C - c \cos B) = b^2 - c^2$.
5. In triangle $ABC$ show that
   $(b+c) \cos A + (c+a) \cos B + (a+b) \cos C = a+b+c$.
6. $LMNO$ is a cyclic quadrilateral in which $MN^2 + ON^2 = OL^2 + LM^2$. Show that $OM$ is a diameter.
7. In triangle $ABC$, $D$ is the point on $BC$ such that $AD = AB$. Show that
$$BD = \frac{c^2 + a^2 - b^2}{a}.$$
8. In triangle $ABC$, $D$ is the mid-point of $BC$. Show that
$$AD^2 = \frac{2b^2 + 2c^2 - a^2}{4}.$$
9. In triangle $ABC$, $D$ is a point on $BC$ such that $BD = ka$, and $E$ is a point on $AC$ such that $CE = kb$.
Show that $DE^2 = k(1-k)c^2 + k(2k-1)b^2 + (1-k)(1-2k)a^2$
10. $ABC$ is a triangle with an obtuse angle at $C$. $CD$ is perpendicular to $AC$, meeting $AB$ at $D$. Show that $CD^2 = \dfrac{AD(BC^2 - DB^2)}{2DB + AD}$.
11. Two lines $AB$ and $CD$ bisect each other. If $BC = a$, $BD = b$, $AB = 2d$ and $CD = 2c$, show that $a^2 + b^2 = 2(c^2 + d^2)$.
12. In triangle $ABC$ show that $\tan C = \dfrac{c \sin A}{b - c \cos A}$. (Hint: draw the perpendicular from $B$.) Hence show that $b \sin c = c \sin(C + A)$. Show also that $\dfrac{b}{\sin A} = c(\cot A + \cot C)$ and $\dfrac{b}{c} \cos A = 1 - \dfrac{a}{c} \cos B$.
13. From the cosine rule show that for any triangle $ABC$
   (a) $bc \cos A + ac \cos B + ab \cos C = \frac{1}{2}(a^2 + b^2 + c^2)$
   (b) $\dfrac{b^2 - a^2}{abc} = \dfrac{\cos A}{a} - \dfrac{\cos B}{b}$

**Exercise B**
1. If in triangle $ABC$ $a = 2x$, $b = 4x$ and $c = 3x$, show that $\sin C = \frac{3}{2} \sin A = \frac{3}{4} \sin B$.
2. If $DA$ bisects angle $A$ of triangle $ABC$ where $D$ lies on $BC$, show that
   (a) $\dfrac{BD}{AB} = \dfrac{DC}{AC}$   (b) $\dfrac{BD}{\sin C} = \dfrac{DC}{\sin B}$
3. If $ABCD$ is a cyclic quadrilateral show that $\dfrac{BD}{\sin A} = \dfrac{AC}{\sin B}$.

4. In triangle $ABC$, $X$ is the mid-point of $BC$. Use the cosine rule to write down two expressions for $AX^2$ and hence show that $b(b - a\cos C) = c(c - a\cos B)$.

5. A perpendicular is drawn from the mid-point of $AB$ in triangle $ABC$ to meet $AC$ at $Y$. Show that $AY = \dfrac{c^2 + b^2 - a^2}{4b}$.

6. $AD$ is the perpendicular from $A$ to $BC$ in triangle $ABC$. Show that $DC = \dfrac{b^2 + a^2 - c^2}{2a}$ and $BD = \dfrac{c^2 + a^2 - b^2}{2a}$. Hence show that $DC - BD = \dfrac{(b-c)(b+c)}{a}$.

7. Sides $BA$ and $CA$ of triangle $ABC$ are produced to $D$ and $E$ so that $AE = \frac{1}{2}AC$ and $AD = BA$. Show that $ED^2 = \frac{1}{4}(2c^2 + 2a^2 - b^2)$.

8. Show that the area of a triangle is given by $\frac{1}{2}bc\sin A$ and also by $\sqrt{s(s-a)(s-b)(s-c)}$. Deduce that $\sin A = \dfrac{2}{bc}\sqrt{s(s-a)(s-b)(s-c)}$.

9. $D$ is the mid-point of $BC$ in triangle $ABC$. Angle $ABD = \theta$. Show that $\tan\theta = \dfrac{2ab\sin c}{b^2 - c^2}$.

10. $ABC$ is a right-angled triangle with angle $A = 90°$. $D$ is the point on $AC$ such that $BD$ bisects angle $B$. Show that $DA = DC\cos B$.

## 7.4 Miscellaneous Problems in Three Dimensions

1. $ABCD$ is the base of a cube. $E, F, G,$ and $H$ are vertically above $A, B, C,$ and $D$ respectively. $I$ is the mid-point of $EF$; $J$ is the mid-point of $GH$. Find
   (a) angle $GAC$.   (b) the angle between planes $AIJD$ and $AEHD$.
   (c) the angle between $AE$ and the plane $AFH$.

2. Find the angle between the faces of a regular tetrahedron.

3. The faces of a wedge are two rectangles $ABCD$ and $ABEF$. Angle $CBE = 60°$, $AB = 10$ cm and $BC = BE = 5$ cm. Calculate angle $DBF$ and the angle between planes $ABCD$ and $CBF$.

4. The height of a right pyramid is 14 cm. It has a square base of side 10 cm. Calculate the angles between
   (a) a face and the base, (b) two opposite faces, (c) two adjacent faces.

5. $ABC$ is a horizontal right-angled triangle with $B = C = 45°$, $AB = a$. $AD$ is a vertical line of length $3a$. Show that if $\theta$ is the angle between face $BCD$ and $ABC$, then $\tan\theta = 3\sqrt{2}$, and that if $\phi$ is angle BDC then $\cos\phi = 0.9$.

6. A rectangular prism has edges $AB$ of length $a$, $AD$ of length $2a$ and $AE$ of length $3a$. If the base is $ABCD$ and corresponding points on the top are $E, F, G, H$, find the angle between plane $BED$ and plane $ABD$.

7. Three vertical lines are cut by a horizontal plane to form a triangle $ABC$. $AB = BC = \frac{3}{4}CA$. $D$ is vertically above $A$ and $AD = AB$. $E$ is vertically above $B$ and $BE = AC$. Find the ratio of the areas of triangles $ABC$ and $CDE$. Hence show that the angle between the planes of the triangles is $\cos^{-1}\left(\dfrac{4}{3\sqrt{5}}\right)$.

8. Three strings of length 2a are connected together at A and to the corners of an equilateral triangle CDE of side 'a', and pulled taut. Show that A is $\sqrt{\frac{11}{3}}a$ above the centroid of the triangle.

9. The bases X, Y, of two poles subtend an angle $2\alpha$ at a point Z on horizontal ground. The tops of the poles A and B subtend angles $\alpha$ and $2\alpha$ at A. Show that if AB is at an angle $\theta$ to the horizontal and $XZ = YZ$ then

$$\tan\theta = \frac{\sec^3\alpha}{2(1 - \tan^2\alpha)}$$

10. A cube has base ABCD and top EFGH. X is a point on AE such that $k \cdot AX = AE$, and Y is a point on CG such that $kGY = GC$. Find angles YXH and HXB.

11. A railway track is in the direction $\theta°$ East of North up a flat incline whose line of greatest slope is at $\alpha$ to the horizontal and in a northerly direction. Show that if $\beta$ is the angle of the railway to the horizontal $\tan\beta = \tan\alpha\cos\theta$, and that if $\gamma$ is the angle between the railway and the line of greatest slope,

$$\sec^2\gamma = \sin^2\alpha + \frac{\cos^2\alpha}{\cos^2\theta}$$

12. An aeroplane is flying directly over a runway at a constant speed of 'a' metres per second. A man stands at a certain position away from the runway and measures the angle of elevation at 0, 1, and 2 seconds to be $\theta_1$, $\theta_2$ and $\theta_3$. If the height of the aeroplane is constant at 100 metres show that the speed of the aeroplane is given by

$$a = \frac{100}{\sqrt{2}}\sqrt{\cot^2\theta_1 + \cot^2\theta_3 - 2\cot^2\theta_2} \text{ metres per second.}$$

13. A hill slopes up at an angle $\alpha$ to the horizontal. Two roads on the hill slope up at $\beta$ and $\gamma$ to the horizontal. If they are at angles $\theta_1$ and $\theta_2$ to a line of greatest slope in the hill where $\sin\theta_1 = 2\cos\theta_2$, show that

$$\sin^2\beta + 2\sin^2\gamma = \sin^2\alpha$$

14. A, B, C are the tops of three posts of heights $a, 2a, 3a$ respectively in a straight line. The posts are distances $a$ and $2a$ apart. The angles of elevation of $A, B, C$ from a point $D$ not in line with the posts are $\alpha, \beta$ and $\gamma$ respectively. Assuming the ground is level show that

$$6 - 2\cot^2\alpha = 9\cot^2\gamma - 12\cot^2\beta$$

15. Suppose two lines are drawn at right angles on the top surface of a wedge whose angle of cross-section is $\theta$. If one line is at an angle $\phi$ to the base show that the other is at an angle

$$\sin^{-1}\sqrt{\sin^2\theta - \sin^2\phi}$$

# 8
# Curves

## 8.1 Notes and Formulae

**Cartesian and Polar Co-ordinates**

*Conversion Formulae*

$$r = \sqrt{x^2 + y^2}, \quad \sin\theta = \frac{y}{\sqrt{x^2 + y^2}}, \quad \cos\theta = \frac{x}{\sqrt{x^2 + y^2}}$$

$$x = r\cos\theta, \quad y = r\sin\theta.$$

The area of a sector between $\theta = \alpha$, and $\theta = \beta$ in polar co-ordinates is $\int_{\alpha}^{\beta} \tfrac{1}{2} r^2 \, d\theta$.

To sketch a curve in polar co-ordinates first make out a table of $r$ against $\theta$ and then plot the corresponding points.

To sketch a curve given in parametric form make out a table of values of $x$ and $y$ for a succession of values of the parameter, and then plot $x$ against $y$.

Do not confuse parametric equations of the form $x = f(\theta)$, $y = g(\theta)$ with polar equations in the form $r = h(\theta)$.

To find the Cartesian equation from the parametric equations, eliminate the parameter between $x$ and $y$.

**Curves of Rational Quadratic Functions**

The following hints will help with sketching curves of the form

$$y = f(x) = \frac{Ax^2 + Bx + C}{ax^2 + bx + c}.$$

Where the numerator and denominator factorise to $\dfrac{A(x-\alpha)(x-\beta)}{a(x-\gamma)(x-\delta)}$ apply some or all of the following tests to gradually build up a picture of the curve:

(a) Apply tests for symmetry $[f(x) = f(-x)]$ or anti-symmetry $[f(x) = -f(-x)]$.

(b) Note that if at least one of $A$ and $a$ is non-zero, then in general the curve gives two values of $x$ for each value of $y$, but if $A$ and $a$ are both zero there is only one value of $x$ for each value of $y$. In both cases the curve is single valued in $y$ (i.e. there is only one value of $y$ for each value of $x$).

(c) Solve $Ax^2 + Bx + C = 0$ to find the zero values.
(d) Solve $ax^2 + bx + c = 0$ to find the asymptotes parallel to $x = 0$.
(e) Find $\lim_{x \to \pm\infty} f(x)$ to find the asymptote parallel to $y = 0$.
(f) Establish where the curve crosses the asymptote parallel to the x-axis (it will *not* cross the other asymptotes).
(g) Establish the sign of the function for values of $x$ between asymptotes and zeros (it will have constant sign in each region) and shade out the prohibited areas.
(h) Establish by differentiation the position and nature of any maxima, minima or points of inflexion.
(i) Obtain the range of possible values of $y$.

For other curves only some of the above techniques may be appropriate. Examination of the differential coefficient to show if the curve is increasing or decreasing and of the second differential coefficient to show if it is convex or concave is often useful.

$$\left( \frac{d^2y}{dx^2} > 0 \text{ if concave upwards}, \quad \frac{d^2y}{dx^2} < 0 \text{ if convex upwards.} \right)$$

## 8.2 Co-ordinates

**Exercise A**
1. Convert the following Cartesian equations to polar form.
   (a) $y = 2x$ (b) $y^2 = 4ax$ (c) $y + 3x = 5$
   (d) $xy = c^2$ (e) $x^4 + y^4 = xy$ (f) $x = 0$
   (g) $(x-a)^2 + (y-b)^2 = 1$
2. Convert the following polar equations to Cartesian form:
   (a) $r = a \sin \theta$ (b) $r = k$ (a constant) (c) $\theta = k$ (a constant)
   (d) $r = a \cos^2 \theta$ (e) $r^2 = a \sin 2\theta$ (f) $r = a(1 - \cos \theta)$.
   (g) $\frac{3}{r} = 1 + 2 \cos \theta$
3. Find polar equations for
   (a) a circle, centre the origin, and radius 'c',
   (b) a straight line of gradient 'm' through the origin.
   (c) a circle of radius 'a' the centre of which is on the initial line and which passes through the origin,
4. Sketch the following curves given in polar form:
   (a) $r = 2 \cos \theta$ (b) $r = \cos 2\theta$ (c) $r = a(1 + \cos \theta)$
   (d) $r = a \sin \frac{1}{2}\theta$ (e) $r^2 = a^2 \sin 2\theta$ (f) $r\theta = 2$
   (g) $r = a \cos^2 \theta$
5. Sketch the following curves given in parametric form:
   (a) $x = t^2$, $y = t$ (b) $x = 1 + t^2$, $y = -t$
   (c) $x = \cos \theta$, $y = \sin \theta$ (d) $x = 3t^2$, $y = 2t^3$
   (e) $x = t^3$, $y = \frac{1}{t^3}$ (f) $x = 2 + 3 \cos \theta$, $y = 3 + 2 \sin \theta$
6. Find the Cartesian equations for
   (a) $x = t^2 + 1$, $y = 2t$ (b) $x = 2 \cos \theta$, $y = 2 \sin \theta$

55

(c) $x = t + \dfrac{1}{t}$, $y = t - \dfrac{1}{t}$   (d) $x = t^2$, $y = \dfrac{1}{t^3}$

(e) $x = \cos 3\theta$, $y = \cos \theta$

7. By putting $x = ty$ find parametric equations for
   (a) $y^2 = 4ax$   (b) $x^2 + y^2 = 9$   (c) $x - y = 1$

**Exercise B**

1. Find the polar equations of
   (a) a circle whose centre is on the initial line at distance '$2a$' from the origin and whose radius is '$a$'.
   (b) a straight line at 45° to the initial line cutting it at a distance '$a$' from the origin.
   (c) a circle which lies above the initial line and touches it at the origin and has radius '$a$'.

2. Sketch the following curves:
   (a) $r = a$   (b) $r = a \sin \theta$   (c) $r = a \sin 3\theta$
   (d) $r^2 = a^2 \cos 2\theta$   (e) $r = a\theta$   (f) $\dfrac{a}{r} = 1 + \cos \theta$
   (g) $r = a \cos \tfrac{1}{2}\theta$

3. Sketch the following curves:
   (a) $x = 1 + t$, $y = 1 - t$   (b) $x = t + \dfrac{1}{t}$, $y = t - \dfrac{1}{t}$
   (c) $x = 2 \sin \theta$, $y = 2 \cos 2\theta$   (d) $x = 2 \sec \theta$, $y = \tan \theta$

4. Find the Cartesian equations of
   (a) $x = \dfrac{1+t}{2-t}$, $y = \dfrac{1-t}{2+t}$   (b) $x = t^2 + 1$, $y = \dfrac{1}{t^4}$
   (c) $x = \sin \theta$, $y = \cos 2\theta$   (d) $x = 3 \sec \theta$, $y = 4 \tan \theta$
   (e) $x = 2 \cos \theta$, $y = 3 \sin \theta$

5. By putting $x = ty$ find the parametric equations for
   (a) $y = mx + c$   (b) $x^2 - y^2 = 1$   (c) $ax^2 + by^2 = y$

## 8.3 Areas in Polar Co-ordinates

1. Find the areas of the following curves between the given values of $\theta$:
   (a) $r = a$ from $\theta = 0$ to $2\pi$   (b) $r = a \cos \theta$ from $\theta = 0$ to $\tfrac{1}{2}\pi$
   (c) $r = a\theta$ from $\theta = \tfrac{1}{4}\pi$ to $\tfrac{1}{2}\pi$   (d) $r = a(2 + \cos \theta)$ from $\theta = 0$ to $\tfrac{1}{2}\pi$
   (e) $r = a \cos 3\theta$ from $\theta = -\tfrac{1}{3}\pi$ to $\tfrac{1}{3}\pi$

2. Find the area of one loop of $r = a \cos 2\theta$.

3. Find the total area of $r = a \sin 2\theta$.

4. In what ratio does $\theta = \tfrac{1}{4}\pi$ divide the area between $\theta = 0$ and $\theta = \tfrac{1}{2}\pi$ of $r = 2\theta$.

5. Sketch and find the area of one loop of
   (a) $r = a \sin \theta$   (b) $r = a \sin 2\theta$   (c) $r = a \sin 3\theta$   (d) $r = a \sin 4\theta$

6. Find the area of one loop of $r^2 = a^2 \sin 2\theta$.

## 8.4 Curve Sketching of Rational Functions
### Exercise A
1. Sketch the following curves showing any asymptotes clearly:
   (a) $y = \dfrac{x-2}{x-1}$
   (b) $y = \dfrac{2x+3}{x+4}$
   (c) $y = \dfrac{x}{(x-2)(x+3)}$
   (d) $y = \dfrac{(x+1)(x-2)}{(2x+1)(x-3)}$
   (e) $y = \dfrac{3x-1}{x^2+3x+2}$
2. Find the ranges of values of y which the following curves *cannot* take for real values of x. (Hint: write the equation as a quadratic in x and write down the condition for x to be imaginary.)
   (a) $y = \dfrac{2}{(x-2)(x+1)}$
   (b) $y = \dfrac{x}{(x+2)^2}$
   (c) $y = \dfrac{2x^2}{x^2-1}$
3. Find the turning points of
   (a) $y = \dfrac{2x}{x^2+1}$
   (b) $y = \dfrac{x^2+2x+1}{x^2+2x+5}$
   (c) $y = \dfrac{2x^2-x-2}{x^2-2x+1}$
   Find the points of inflexion in (a).
4. State whether the following curves are symmetrical, anti-symmetrical, or neither. If symmetrical give the axes of symmetry.
   (a) $y = x^2$
   (b) $x^2 + y^2 = 2x$
   (c) $y = x^3 - 1$
   (d) $xy = (x-1)^2$;
   (e) $(x+y)^4 = 2x^2 + y^3$;
   (f) $(xy)^2 = 2x - 1$;

### Exercise B
1. Sketch the following curves showing any asymptotes clearly:
   (a) $y = \dfrac{(x-2)(x+4)}{(2x+1)(3x-2)}$
   (b) $y = \dfrac{(x-3)^2}{(x+3)^2}$
   (c) $y = \dfrac{2x^2-x-1}{x^2+x-6}$
2. Find the range of values of y which the following curves *cannot* take:
   (a) $y = \dfrac{3}{(x-1)^2}$
   (b) $y = \dfrac{x^2}{2x+3}$
   (c) $y = \dfrac{x^2+2}{(x-3)(2x+1)}$
3. Find the turning points of
   (a) $y = \dfrac{x^2}{x^2+2x-1}$
   (b) $y = \dfrac{x^2-2x}{x^2-2x-2}$
   (c) $y = \dfrac{(x+1)^2}{(x-1)^2}$
4. Find if the following curves are symmetrical, anti-symmetrical or neither. State any axes of symmetry or antisymmetry.
   (a) $xy = 1$
   (b) $x^2 = y + x$
   (c) $x = 4y^2$
   (d) $y = \sin(x^3)$
   (e) $(y+x)^2 = 1$
   (f) $y = (x-1)^2$

## 8.5 Miscellaneous Curves
1. Sketch the following curves:
   (a) $y = 2(x-1)^3$
   (b) $y = 2x(x-1)^2$
   (c) $y = \dfrac{2x}{(x-1)^2}$
2. Find the stationary points and their nature of $y = \dfrac{x^3+2}{3x}$. Sketch the curve $y = \dfrac{3x}{x^3+2}$.

3. Find the turning points and the nature of $y = x(x-4)^4$. Sketch the curve.
4. Sketch $y = x + \dfrac{2}{x} + 3$ and find the turning points.
5. Determine the nature and position of the stationary points of $y = x^5 - x$. Sketch the curve.
6. Find the asymptotes and zeros of $y = 4 + \dfrac{2}{x+1}$. Sketch the curve.
7. Sketch $y = \dfrac{2}{1+2x^2}$ and $y = 2x+1$. Hence find an approximate solution to $4x^3 + 2x^2 + 2x - 1 = 0$ near $x = 0$.
8. Sketch $y = \dfrac{1}{(2+x)^2}$ and $y = x$ on the same axes. By redrawing a suitable section between $x = 0$ and $x = 1$ find a solution to $x^3 + 4x^2 + 4x = 1$ accurate to 2 decimal places.
9. Solve the following simultaneous equations graphically:
   (a) $y = x + \dfrac{1}{x}$, $y = 2x^2$    (b) $y = 3x$, $y = \dfrac{4}{(x-1)^2}$ for $1 < x < 2$.
   (c) $y = x(x-1)^2$, $y = \dfrac{2}{(x+1)^2}$ for $1 < x < 2$.
10. Sketch the following curves:
    (a) $y = 2^x$    (b) $y = 3^x$    (c) $y = 3^{x+2}$
    (d) $y = e^x$    (e) $y = e^{-x}$    (f) $y = \tfrac{1}{2}(e^x + e^{-x})$
    (g) $y = \tfrac{1}{2}(e^x - e^{-x})$
11. Sketch (a) $y = \log_{10} x$    (b) $y = \log_e x$    (c) $y = \log_{10}(1+x)$
             (d) $y = \log_e (3x)$    (e) $y = \log_e (5x-2)$

# 9

# Inequalities

## 9.1 Notes and Formulae

An inequality remains true if it is multiplied or divided by a *positive* quantity or added to or subtracted from by *any* quantity.

The inequality must be reversed if multiplied or divided by a *negative* quantity. If the reciprocal of both sides is taken the inequality is reversed for *positive* numbers on both sides; but cases involving negative numbers must be carefully checked. We must be careful to check the validity of any result when we apply the same operation to both sides of an inequality.

The use of graphs is useful in finding solutions to inequalities. Solutions are found from intersecting regions between the lines or curves.

It is often convenient to consider cases where a variable takes positive values first, and then negative values.

e.g. to find where $x - 3 > \dfrac{4}{x}$

First consider $x$ *positive* giving $\quad x(x-3) > 4$
or $\quad\quad\quad\quad\quad\quad\quad\quad\quad\quad\quad (x-4)(x+1) > 0$
or $\quad\quad\quad\quad\quad\quad\quad\quad\quad\quad\quad x > 4$ (reject $x < -1$).
Then taking $x$ *negative* giving $\quad x(x-3) < 4$
or $\quad\quad\quad\quad\quad\quad\quad\quad\quad\quad\quad (x-4)(x+1) < 0$
or $\quad\quad\quad\quad\quad\quad\quad\quad\quad\quad\quad -1 < x < 0$ (reject $0 < x < 4$).
The solution is then $x > 4$ or $-1 < x < 0$.

$|x|$ means the numerical value of $x$, e.g. $|-2| = 2$,

$|x| < 1$ therefore means $-1 < x < 1$.

For the quadratic function $f(x) \equiv ax^2 + bx + c$, we can show that the function takes the same sign as '$a$' except where the roots of $f(x) = 0$ are real and unequal and $x$ lies between them.

To establish the sign of a function of the form $\dfrac{f(x)}{g(x)}$ where $f(x)$ and $g(x)$ can be factorised into linear factors, split the range of $x$ into intervals whose boundaries are the solutions of $f(x) = 0$ and $g(x) = 0$, and choose a representative value of $x$ in each region. All the values of $x$ in each region will give the same sign to the function as this representative value.

To establish minimum or maximum values of a quadratic function complete the square in the $x$ terms.

e.g. $x^2 + 2x + 5 = (x+1)^2 + 4$, hence minimum value is 4 when $x = -1$.
and $1 + 3x - x^2 = 3\tfrac{1}{4} - (x - \tfrac{3}{2})^2$, hence maximum value is $3\tfrac{1}{4}$ when $x = 1\tfrac{1}{2}$.

In straightforward cases where regions are to be plotted from inequalities, first plot the corresponding **equality** and then establish which side of the line is required by checking just one point.

e.g. to find the region satisfying $x^2 + y^2 < 1$, plot $x^2 + y^2 = 1$. Use (say) $(0, 0)$ as a check point. This satisfies the inequality so the region is inside the circle.

In plotting regions which obey more than one inequality, plot each region separately and shade *out* the region *not* required. The final unshaded region satisfies all of the inequalities. Be careful to emphasise where lines are included in the required regions.

## 9.2 Basic Inequalities

1. Establish which values of $x$ satisfy

  (a) $2x + 1 < 3x - 5$    (b) $-5x - 3 < x + 4$    (c) $\dfrac{1}{x} < -2$

(d) $7x - 5 \geq 5 - 8x$  (e) $\dfrac{1}{x} \leq x$  (f) $\dfrac{2}{3+x} < 4$

(g) $\dfrac{2x+1}{x-1} \leq 6$  (h) $\dfrac{2}{x+1} > \dfrac{3}{x-1}$  (i) $\dfrac{4x+3}{2x-1} \geq 2$

2. Sketch graphs to establish the values of $x$ which make the following true:
 (a) $x^2 - 1 > 0$  (b) $x^3 - 4x - 5 < 0$  (c) $3x^2 - 1 < 2x$
 (d) $2x^2 \leq 3 + 2x$  (e) $4x + 2 > \dfrac{1}{x}$  (f) $3x^2 + 1 < 2x - 2$
 (g) $\dfrac{x-1}{x+1} > \dfrac{x+3}{x-2}$  (h) $x^2 + 3 > \sin x$

3. Find the values of $x$ which satisfy
 (a) $|x-2| > 3$  (b) $|x+1| < |x|$;
 (c) $|x-1| < |x-2|$  (d) $|x+1| > 4|x-3|$
 (e) $2|x-2| < 3|2x+5|$  (f) $|1-x| > |x+2|$
 (g) $-1 < x < 2$ and $1 < x < 3$  (h) $|x| \leq 3$ and $1 < x < 5$
 (i) $|2x+4| > 3$ and $|x| > 0$

4. Plot the following inequalities. Shade out the region *not* required leaving the region which satisfies the inequality (or inequalities) unshaded. Show clearly if the points on the boundary lines are included.
 (a) $x^2 + y^2 < 3$  (b) $y > x$
 (c) $y^2 \leq 2x$  (d) $\dfrac{x^2}{4} + \dfrac{y^2}{9} \leq 1$
 (e) $y < \log_e x$  (f) $2y - 3 \geq e^x$
 (g) $y > 2x + 1$, $x < 2$  (h) $x^2 > y$, $y < 2x + 3$
 (i) $x^2 + 2y^2 < 3$, $x^2 + y^2 > \dfrac{1}{4}$  (j) $y > x + 1$, $(y-2)^2 < 4x$, $y > -x$
 (k) $x < \log_e y$, $y \leq \dfrac{1}{x}$, $y > -1$  (l) $|x-1| < |x-2|$
 (m) $4|x| > |x+3|$  (n) $-\pi < x < \pi$, $y \geq \sin x$, $y \leq \cos x$

## 9.3 Quadratic Inequalities

**Exercise A**
1. What values of $x$ make the following functions positive?
 (a) $(x-3)(x+3)$  (b) $(x-2)(x+4)$  (c) $(2x-1)(x+2)$
 (d) $-(3x-5)(2x-7)$  (e) $(5x+7)(7x+5)$
2. Find the values of $x$ which make the following functions positive:
 (a) $x^2 + x - 2$  (b) $2x^2 - 5x + 3$  (c) $-x^2 + x + 2$
 (d) $6x^2 - x + 1$  (e) $6x^2 - x - 2$
3. Find the values of $k$ which make the following inequalities true for all $x$:
 (a) $x^2 - 2x + k > 0$  (b) $2x^2 - x + k > 2$

(c) $kx^2 - 2x + 1 > 0$      (d) $6x^2 + 3x + k - 2 > 6$
(e) $-(k+1)x^2 - 2x + 1 < 2$      (f) $-x^2 + 3x - 4 < k$
4. By completing the square show that the following inequalities are true for all $x$, and establish the minimum or maximum values of the functions involved.
(a) $x^2 + 2x + 8 > 0$      (b) $-2x^2 + 3x - 2 < 0$      (c) $3x^2 - x + 1 > 0$
(d) $-2 - 5x^2 + 6x < 0$      (e) $-1 - 3x - 6x^2 < 0$
5. Establish the values of $x$ which satisfy
(a) $(x-1)(x+2)(x+3) > 0$      (b) $(x+4)(x-7)(x-8) < 0$
(c) $\dfrac{(x+2)(x-1)}{x-3} \leqslant 0$      (d) $(x-3)(x+4)(x-5) \geqslant 0$
(e) $\dfrac{(x+1)(x-2)}{x(2x+3)} < 0$      (f) $\dfrac{(x-3)(2x+5)}{(x+7)(x-6)} \geqslant 0$

## Exercise B

1. What values of $x$ make the following functions negative?
(a) $(x+4)(x-4)$    (b) $(2x-3)(x-2)$    (c) $-3(x+5)(4x-2)$
2. Find the values of $x$ which make the following functions negative:
(a) $x^2 - 1$    (b) $2x^2 + 5x - 3$    (c) $-2x^2 + x + 6$
3. Find the values of $k$ which make the following inequalities true for all values of $x$:
(a) $kx^2 + x - 2 > 0$      (b) $kx^2 + 3x - 1 < 3$
(c) $2x^2 + x - k > 4$      (d) $kx^2 + 2 > 3x$
(e) $2x^2 + kx + 2 > 1$      (f) $k^2x^2 - 2(k+1)x + 1 > 0$
4. By completing the square establish whether the following inequalities are true for all values of $x$, and find the values of $x$ which make the functions maximum or minimum, stating which.
(a) $x^2 - 2x + 4 > 0$      (b) $-2x^2 + x + 1 < 0$
(c) $1 + 3x - 4x^2 < 0$      (d) $(2x+1)^2 - x > 0$
(e) $(2x-1)^2 - (3x+2)^2 + 6x + 7 < 0$
5. Find the values of $x$ which satisfy
(a) $(x-1)(x+3)(x+6) > 0$      (b) $(x+1)(2x-3)(3x+4) \leqslant 0$
(c) $(2x+1)(3x-2)(4x+3) \geqslant 0$      (d) $\dfrac{(x+1)(x-3)}{3x+2} \leqslant 0$
(e) $\dfrac{2x^2 - 3x - 2}{x^2 + x} > 0$      (f) $x^3 + 3x^2 - x < 3$
(g) $\dfrac{3x^2 + 7x + 2}{6x^2 - 7x + 2} \leqslant 0$

## 9.4 Miscellaneous

1. Find the values of $k$ which make the following inequalities true for all $x$:
(a) $x^2 - 2(k-1)x + 1 > 0$      (b) $x^2 + kx > 2k$
(c) $kx^2 + (k-1)x + k > 0$      (d) $2(k+2)x^2 + 3kx + k > 1$

2. Find the values of $k$ which make the following true for all $x$, and find the value of $x$ which makes the function a minimum, and the value of $k$ which makes this minimum value zero.
   (a) $x^2 + 2x + k \geq 0$
   (b) $x^2 - 3x + k - 1 \geq 0$
   (c) $4x^2 + 2x + k^2 \geq 0$
   (d) $4x^2 + 2(k+1)x + 2k - 1 \geq 0$
3. If $x < y$ establish whether the following are *always* true:
   (a) $\log x < \log y$
   (b) $e^x < e^y$
   (c) $\sin x < \sin y$
4. Find the values of $x$ which satisfy
   (a) $\dfrac{1}{x-2} > \dfrac{1}{x-3}$
   (b) $\dfrac{1}{x} > \dfrac{x}{x-3}$
   (c) $\dfrac{1}{x-1} > \dfrac{x}{x-2}$
   (d) $\dfrac{x}{x-3} > \dfrac{1}{x+1}$
5. Find the values of $x$ which satisfy
   (a) $(x+2)(x+1)(x+3) > 2(x+2)$
   (b) $(x+1)(x+2) < \dfrac{1}{(x+1)(x+2)}$
   (c) $x(x-1)(x-4) < (x-2)(x-3)(x-5)$

   Hint: Find where the left and right hand sides are positive or negative to give approximate graphs and find the points of intersection of the curves.
6. Find the values of $x$ which satisfy
   (a) $\left|\dfrac{x-2}{x+1}\right| < \left|\dfrac{x+3}{x-4}\right|$
   (b) $|(x+1)(x+3)| > |x+4||x-2|$

# 10

## Co-ordinate Geometry

### 10.1 Notes and Formulae

**Points**
If $P$ is the point $(x_1, y_1)$ and $Q$ is the point $(x_2, y_2)$
(a) the distance $PQ$ is $\sqrt{(x_1 - x_2)^2 + (y_1 - y_2)^2}$,
(b) mid-point of $PQ$ is $\{\frac{1}{2}(x_1 + x_2), \frac{1}{2}(y_1 + y_2)\}$.
(c) If $R$ divides $PQ$ in the ratio $l:m$, $R$ is the point
$$\dfrac{lx_2 + mx_1}{l+m}, \dfrac{ly_2 + my_1}{l+m}$$

(d) The gradient of $PQ$ is $\dfrac{y_2 - y_1}{x_2 - x_1}$.

## Straight Lines

Two lines $y = m_1 x + c_1$ and $y = m_2 x + c_2$ are parallel if $m_1 = m_2$.
They are perpendicular if $m_1 m_2 = -1$.
The straight line of gradient '$m$', and with intercept '$c$' on the $y$-axis, is $y = mx + c$.
The line through the point $(x_1, y_1)$ and with gradient '$m$' is $y - y_1 = m(x - x_1)$.
The straight line joining $(x_1, y_1)$ and $(x_2, y_2)$ is

$$y - y_1 = \frac{y_2 - y_1}{x_2 - x_1}(x - x_1) \quad \text{or} \quad \frac{y - y_1}{y_2 - y_1} = \frac{x - x_1}{x_2 - x_1}.$$

If a line cuts the $x$-axis at $(a, 0)$ and the $y$-axis at $(0, b)$ then its equation is

$$\frac{x}{a} + \frac{y}{b} = 1$$

If the perpendicular from the origin to a straight line is of length '$p$' and at an angle $\theta$ to the $x$-axis, then the line has the equation

$$x \cos \theta + y \sin \theta = p$$

The length of the perpendicular from the point $(x_1, y_1)$ to the straight line $ax + by + c = 0$ is

$$\pm \frac{ax_1 + by_1 + c}{\sqrt{a^2 + b^2}}$$

the sign being chosen to give a positive length.
If $y = m_1 x + c_1$ and $y = m_2 x + c_2$ cut at an angle $\theta$ then

$$\tan \theta = \frac{m_1 - m_2}{1 + m_1 m_2}$$

## Loci and Tangents

The equation of a locus is a relationship between the co-ordinates of a point on the locus which is true for all points on the locus, and conversely any point which obeys the equation must be on the locus.

To find the tangent to a curve at the point $(x_1, y_1)$, calculate the value of $\dfrac{dy}{dx}$ at $(x_1, y_1)$ and use the equation $y - y_1 = m(x - x_1)$.
If the gradient of the tangent is '$m$', the gradient of the normal is $-\dfrac{1}{m}$.
In the case of parametric equations whose parameter is '$t$'

$$\frac{dy}{dx} = \frac{dy}{dt} \times \frac{dt}{dx} = \frac{\frac{dy}{dt}}{\frac{dx}{dt}}$$

## The Circle
The circle, centre the origin, radius 'r', is $x^2 + y^2 = r^2$.
The circle, centre $(a, b)$, radius 'r', is $(x-a)^2 + (y-b)^2 = r^2$.
If $g^2 + f^2 - c > 0$, then $x^2 + y^2 + 2gx + 2fy + c = 0$ is a circle with centre $(-g, -f)$ and radius $\sqrt{g^2 + f^2 - c}$.
The circle on $AB$ as diameter where $A$ is $(x_1, y_1)$ and $B$ is $(x_2, y_2)$ is
$$(x - x_1)(x - x_2) + (y - y_1)(y - y_2) = 0$$
The tangent to the circle $x^2 + y^2 + 2gx + 2fy + c = 0$ at $(x_1, y_1)$ on its circumference is
$$xx_1 + yy_1 + g(x + x_1) + f(y + y_1) + c = 0$$
If $P(x_1, y_1)$, is outside the circle the length of the tangent to $x^2 + y^2 + 2gx + 2fy + c = 0$ is
$$\sqrt{x_1^2 + y_1^2 + 2gx_1 + 2fy_1 + c}$$
If $O_1 = x^2 + y^2 + 2g_1 x + 2f_1 y + c_1$ and $O_2 = x^2 + y^2 + 2g_2 x + 2f_2 y + c_2$ and $O_1 = 0$ and $O_2 = 0$ cut at $A$ and $B$, then $O_1 + \lambda O_2 = 0$ is a family of circles through $A$ and $B$ (where $\lambda$ takes any real value). $O_1 - O_2 = 0$ is the common chord through $AB$. $O_1 = 0$, and $O_2 = 0$ cut orthogonally (i.e. at right angles) if $2g_1 g_2 + 2f_1 f_2 = c_1 + c_2$.

Questions on circles are often simplified by using a clear diagram and making use of geometrical properties.

To find the equation of a particular circle we require three constants, $g$, $f$, and $c$. Forming three simultaneous equations from the given conditions will lead to the solution.

## The Parabola
The locus of a point whose distance from a fixed point, the focus, is equal to its distance from a fixed straight line, the directrix, is a parabola. If the focus is $(a, 0)$ and the directrix is $x = -a$, the parabola is $y^2 = 4ax$. The parametric equations are $x = at^2$, $y = 2at$.

The point where the line of symmetry cuts the parabola is called the vertex. If the vertex is $(h, k)$ instead of $(0, 0)$ the parabola is $(y - k)^2 = 4a(x - h)$.

The tangent at $(x_1, y_1)$ on the parabola $y^2 = 4ax$ is $yy_1 = 2a(x + x_1)$.

The gradient at the point with parameter 't' is $\frac{1}{t}$ and the tangent is $ty = x + at^2$.

The normal is $y + tx = 2at + at^3$.

The line $y = mx + c$ is a tangent to $y^2 = 4ax$ if $c = \frac{a}{m}$.

In tackling problems on the general properties of a parabola it is wise to *choose* the axes to make the equation $y^2 = 4ax$ and then use the parametric form $x = at^2$, $y = 2at$ which usually simplifies the work.

## Conic Sections in General
Cross sections of a cone may take a variety of forms depending on where the plane cuts the cone. The form of the equation of the curve is in general a

quadratic and can take the following forms for a suitable choice of axes:

| | |
|---|---|
| $(a_1x + b_1y + c_1)(a_2x + b_2y + c_2) = 0$ | Pair of straight lines |
| $x^2 + y^2 = a^2$ | Circle |
| $y^2 = 4ax$ | Parabola |
| $\dfrac{x^2}{a^2} + \dfrac{y^2}{b^2} = 1$ | Ellipse |
| $\dfrac{x^2}{a^2} - \dfrac{y^2}{b^2} = 1$ | Hyperbola |

The parabola, ellipse and hyperbola fit similar definitions: the locus of a point whose distance from a fixed point (the **focus**) is $e$ times its distance from a fixed straight line (the **directrix**). $e$ is a constant called the **eccentricity**.

If $e = 1$ we have a parabola.
If $e < 1$ we have an ellipse.
If $e > 1$ we have a hyperbola.

For the ellipse, $b^2 = a^2(1 - e^2)$ and $2a$ and $2b$ are the lengths of the major and minor axes.
For the hyperbola $b^2 = a^2(e^2 - 1)$.

Work for the ellipse and hyperbola is in general very similar. e.g. the tangent at $(x_1, y_1)$ to the ellipse is $\dfrac{xx_1}{a^2} + \dfrac{yy_1}{b^2} = 1$ and to the hyperbola is $\dfrac{xx_1}{a^2} - \dfrac{yy_1}{b^2} = 1$. Both ellipse and hyperbola have foci $(\pm ae, 0)$ and directrices $x = \pm \dfrac{a}{e}$.

The *latus rectum* of a conic is the chord through the focus perpendicular to the axis of symmetry.

Parametric equations:

Ellipse $\quad x = a\cos\theta,\ y = b\sin\theta$
Hyperbola $\quad x = a\sec\theta,\ y = b\tan\theta$

For those familiar with hyperbolic functions, alternative equations for the hyperbola are

$$x = a\cosh u,\ y = b\sinh u$$

If the locus of mid-points of chords of an ellipse parallel to $y = mx$ is $y = m_1x$, then $y = mx$ is the locus of mid-points of chords parallel to $y = m_1x$ and $mm_1 = -\dfrac{b^2}{a^2}$. These are called **conjugate diameters**.

$y = mx + c$ is a tangent to
(a) the ellipse if and only if $c^2 = a^2m^2 + b^2$.
(b) the hyperbola if and only if $c^2 = a^2m^2 - b^2$.

For the ellipse, in the parametric equations $x = a\cos\theta,\ y = b\sin\theta$, $\theta$ is called the **eccentric angle**.
$x^2 + y^2 = a^2$ is called the **auxilliary circle**.

At $(a\cos\theta, b\sin\theta)$ the tangent to the ellipse is $bx\cos\theta + ay\sin\theta = ab$ and the normal is $ax\sin\theta - by\cos\theta = (a^2 - b^2)\sin\theta\cos\theta$.

At $(a\sec\theta, b\tan\theta)$ the tangent to the hyperbola $bx - ay\sin\theta - ab\cos\theta = 0$. The normal is $ax\sin\theta + by = (a^2 + b^2)\tan\theta$.

The hyperbola $\dfrac{x^2}{a^2} - \dfrac{y^2}{b^2} = 1$ has asymptotes $y = \pm\dfrac{b}{a}x$.

If the asymptotes are at right angles we have a **rectangular hyperbola**. The asymptotes can be used as axes and the equation becomes $xy = c^2$ where $c^2 = \frac{1}{2}a^2$.

Work on the rectangular hyperbola is simplified by using the parameters $x = ct$, $y = \dfrac{c}{t}$. The tangent at $\left(ct, \dfrac{c}{t}\right)$ is $x + t^2 y - 2ct = 0$.

The normal is $t^2 x - y - ct^3 + \dfrac{c}{t} = 0$.

**Change of Axes**

If new axes are taken at $(x_1, y_1)$ parallel to the original axes of a curve, the new equation is found by putting $x + x_1$ for $x$ and $y + y_1$ for $y$.

## 10.2 The Straight Line

**Exercise A**

1. Find the lengths of the lines joining
   (a) (1, 2) and (4, 6)
   (b) (10, −1) and (2, −16)
   (c) (−4, −3) and (−5, −12)
   (d) (0, 0) and (7.1, −3.4)
   (e) $(a, 2a)$ and $(-a, -2a)$
   (f) $\left(t, \dfrac{1}{t}\right)$ and $\left(\dfrac{1}{t}, t\right)$
   (g) $(1 + 2t, 3t - 1)$ and $(t + 3, 2t - 3)$
2. Find the mid-points of the lines in question 1.
3. If $A$ is $(-2, 3)$ and $B$ is $(4, 8)$, find $C$ such that $AC = 5CB$.
4. Find the points $L, M$ which divide the line joining $(a, 2a)$ and $(10a, 8a)$ into three equal parts.
5. Find the gradients of the lines in question 1.
6. Find the gradients of the lines perpendicular to those in question 1.
7. Show that A (2, 4), B (5, 3), C (2, 2) and D (−1, 3) form the vertices of a parallelogram. Find the point of intersection of the diagonals.
8. Show that (4, 6), (5, 5) and (−1, 1) lie on a circle centre (2, 3). What is its radius?
9. Find the equations of the following lines:
   (a) With gradient 6 and passing through (0, 3).
   (b) With gradient 4 through the origin.
   (c) With gradient −1 through (−1, 2).
   (d) Joining (3, 5) and (9, 7).
   (e) Parallel to $6y - 4x = 7$ and passing through $(-2, -4)$.
   (f) Perpendicular to $x + 3y = 4$ and passing through the point of intersection of $y = x$ and $2y + 3 = 0$.

(g) Cutting the y-axis at (0, −1) and the x-axis at (2, 0).
  (h) Bisecting the line joining (4, 7) and (9, −1) at right angles.
  (i) At an angle of 30° to the x-axis and bisecting the line joining (0, 2) and (−3, 5).
  (j) The line whose perpendicular from the origin is of length 2 and is at 20° to the y-axis.
  (k) The line at an angle of 75° to the x-axis and distant 3.4 from the origin.
10. Find the distance of
  (a) $(3, -1)$ from $4x - 3y = 6$.
  (b) $(7, 5)$ from $4y = 7x - 2$.
  (c) $(t, 3t)$ from $5x + 12y = 13$.
  (d) $(a^2b, b^2a)$ from $a(y + a) = b(x + b)$
11. Find the acute angles between
  (a) $y = 7x + 1$ and $4y = 7x - 3$
  (b) $2.4x = 3.4y + 1.3$ and $y = 2.4x$
  (c) $5x = 12y - 1$ and $5x + 12y + 3 = 0$

# Exercise B

1. If $A$ is $(1, 10)$ and $B$ is $(-17, 21)$ and $C$ is the mid-point of $AB$, find two possible points $D$ such that $3CD = DB$.
2. Show that $A(1, 1)$, $B(-2, 4)$, and $C(4, 4)$ form a right-angled isosceles triangle. If the point $D$ is such that $ABDC$ is a square, find the co-ordinates of $D$.
3. If $(a, 7)$, $(3, 1)$ and $(-1, 2)$ are the vertices of an isosceles triangle, find $a$. Find the mid-point of the base and the co-ordinates of a fourth point which completes a rhombus. Show that the diagonals of the rhombus are perpendicular to each other.
4. Find the equations of the straight line:
  (a) of gradient $-2$, cutting the y-axis at $(0, -3)$.
  (b) of gradient $-\frac{3}{4}$ cutting the x-axis at $(3, 0)$.
  (c) parallel to $2y - 3x = 6$ and through the origin.
  (d) perpendicular to $5x - 7y + 6 = 0$ and through $(7, -2)$.
  (e) perpendicular to the line joining $(-2, 3)$ and $(4, -8)$ and through the mid-point of the line joining $(6, 2)$ and $(6, -1)$.
  (f) through the points of intersection of $y = 3$ and $x = 2$ and of $x = 3$, and $y = 2$.
  (g) parallel to $x = 0$ so that the triangle formed by itself, $y = 0$ and $3x = 2y + 3$ has an area of 10 square units.
  (h) parallel to and distant two units from $y + 7 = 2x$.
  (i) that bisects the acute angle between $2.4x + 0.7y = 10.3$ and $1.5x - 3.6y = 4.9$.
  (j) that together with $y = 12$ and $\sqrt{3}y = 3x + \sqrt{3}$ forms an equilateral triangle such that the line passes through $(\frac{1}{2}\sqrt{3}, 0)$.
  (k) whose distance from the origin is equal to the length of the line joining $(2, 5)$ and $(-7, 5)$ and is perpendicular to $y + 4 - 7x = 0$.
5. Find the distance of
  (a) $(3, -2)$ from $y = 4x - 3$
  (b) $(12, 5)$ from $2y = 3x - 1$
  (c) $(2, -1)$ from $5x = 2y + 12$

(d) $(a, 0)$ from $y = 3x + 4a$
6. Find the equation of the line bisecting the obtuse angle between
   (a) $4y = 3x - 2$ and $y = x + 4$
   (b) $y + x = 12$ and $y = 6 - 3x$
   (c) $2y + x = 7$ and $-4y = x - 3$

## 10.3 Loci

**Exercise A**
1. Show that the given points are on the following curves:
   (a) $(1, 5)$ on $y = 3x + 2$ (b) $(2, 7)$ on $y = 2x^2 - 1$
   (c) $(m, m^2 + c)$ on $y = mx + c$ (d) $(t^2, 2t)$ on $y^2 = 4x$
   (e) $(4\frac{1}{2}, 1\frac{1}{2})$ on $x^2 - y^2 = 18$ and on $x + y = 6$
2. Find the points of intersection of
   (a) $y = 3x - 1$ and $y = 2x + 4$ (b) $y^2 = 2x$ and $y = 4$
   (c) $y^2 + x^2 = 9$ and $x + y = 2$ (d) $xy = 22$ and $4y = x - 8$
   (e) $\dfrac{x^2}{25} + \dfrac{y^2}{16} = 9$ and $(x - 3)^2 + (y - 2)^2 = 9$
3. In the following equations eliminate the parameter to find the Cartesian equation of the curve:
   (a) $x = t, y = t + 1$ (b) $x = t^2 + 1, y = 3t + 4$
   (c) $x = \sin t, y = \cos t$ (d) $x = at^2, y = 2at$
   (e) $x = 4 \sec t, y = 3 \tan t$ (f) $x = \sqrt{t - 1}, y = t^2$
4. Plot the following curves:
   (a) $x = t + 2, y = t^2$ from $t = -2$ to $t = 3$
   (b) $x = t - 1, y = 2t + 4$ from $t = 0$ to $t = 6$
   (c) $x = t^2, y = \dfrac{1}{t + 1}$ from $t = 0$ to $t = 5$
   (d) $x = \dfrac{3 + t}{1 + 2t}, y = \dfrac{5 - t}{4t - 1}$ from $t = 1$ to $t = 8$
   (e) $x = 2 \tan \theta, y = 3 \sec \theta$ from $\theta = 0°$ to $\theta = 80°$
5. Find the values of the parameter $t$ in the following cases:
   (a) Where $x = 2t, y = t$ meets $y^2 - 3 = x$
   (b) Where $x = t, y = \dfrac{1}{t}$ is distant 1 unit from $3x + 4y = 2$
   (c) Where the line joining $(1, 0)$ to $(4, 4)$ meets $x = t^2, y = 2t$.
6. Find the tangents and normals to
   (a) $y = 4x^2 + 2$ at $(2, 18)$ (b) $y^2 = 2x - 3$ at $(2, 1)$
   (c) $xy + x^2 + y^2 = 5$ at $(0, \sqrt{5})$ (d) $x^3 + 2y^3 = x - 10$ at $(2, -2)$
   (e) $x = t, y = \dfrac{1}{t}$ at $t = 2$ (f) $x = \cos t, y = t + \sin t$ at $t = \dfrac{1}{4}\pi$
   (g) $x = a \sec t, y = b \tan t$ at any point '$t$'

7. Find the tangent to $2x^2 + 3y^2 = 1$ which passes through (1, 2).
8. Find the value of $t$ for which the normal to $x = t^3 + 1$, $y = \frac{1}{t^2}$ is parallel to the tangent to $y^2 = 8x$ at (2, 4).
9. Find the points at which the tangent to $x = 3t^2$, $y = 6t$ at $t = 2$ cuts the co-ordinate axes and the area between the axes and the tangent.
10. $P$ is the point $(h, k)$. Find the relationship which must hold between $h$ and $k$ if
    (a) $P$ is equidistant from $x = 0$ and $y = 0$.
    (b) $P$ is distant 8 units from $8x + 15y = 1$.
    (c) $P$ is the same distance from $x = -a$ as from $(a, 0)$
    (d) the distance of $P$ from $(a, 2a)$ is a quarter of its distance from $(2a, a)$.
    (e) the perpendicular from $P$ to the tangent at (4, 4) to $y^2 = 4x$ is of length 2 units.
    (f) $A$ is (2, 3), $B$ is (5, 7) and triangle $APB$ has area 12 units
    (g) the line from $(-2, 3)$ to $P$ is perpendicular to the tangent at '$t$' to $x = at^2$, $y = 2at$.
11. Find the locus of
    (a) the point which is equidistant from (2, 1) and $(-7, -5)$
    (b) $P$, such that $AP^2 + BP^2 = 10$, where $A$ is (5, 7) and $B$ is (6, 5).
    (c) $P$, such that $AP - BP = 3$, where $A$ is (1, 0) and $B$ is (0, 1).
    (d) the mid-point of $PQ$ where $P$ and $Q$ are the points of intersection with the $x$ and $y$ axes of the tangent to $x = t^3$, $y = t^2$ at any point.
    (e) the point $P$ such that the angle $APB = 90°$ where $A$ is (3, 1) and $B$ is $(-2, -6)$.
    (f) the mid-point of $AB$ where $B$ is (1, 0) and $A$ is any point on the normal to $y^2 + (x+2)^2 = 4$ at $(-1, \sqrt{3})$.

## Exercise B
1. Find the value of the parameter $\theta$ in the following cases:
    (a) where $x = \cos\theta$, $y = \cos 2\theta$ meets $x = y$, for $0 \leq \theta \leq \pi$.
    (b) where $x = 2\cos\theta$, $y = 3\sin\theta$ is at a distance of $\sqrt{5}$ units from the origin, for $0 \leq \theta \leq 360°$.
    (c) where $x = \theta\cos\theta$, $y = \theta\sin\theta$ cuts the circle $x^2 + y^2 = 9$.
2. Find tangents and normals to
    (a) $y = \sqrt{3x+1}$ at (8, 5).
    (b) $2y^2 + xy = 6$ where the curve cuts $x = y$ in the region $x > 0$.
    (c) $x = t^2$, $y = t^3$ at $t = 3$.   (d) $x = \frac{1+t}{1-t}$, $y = 1 - t^2$ at $t = 4$.
    (e) $x = a\cos\theta$, $y = b\sin\theta$ where the curve is parallel to $x + y = 1$, $\theta$ being acute and $a$ and $b$ both being positive.
    (f) $x = 2t$, $y = \frac{2}{t}$ where the curve meets $2x - y = 2$.
3. Find the point of intersection of the normal to $x^2 + y^2 + 2x = 39$ at (5, 2) and the tangent to $x = \frac{1}{t+2}$, $y = (t-1)^2$ where $t = 4$.
4. Find the equation of the normal from (0, 4) to $x = t^3$, $y = t^2$.

5. Find the distance from (1, 6) to the tangent to $x = (t-1)^2$, $y = 2(t+2)$ at the point with parameter '$p$'. Hence find the value of $p$ for which this is a minimum.
6. $P$ is the point $(h, k)$. In each of the following cases find the relationship between $h$ and $k$:
   (a) The distance of $P$ from the origin is 6 units.
   (b) $P$ is the same distance from $2x + 3y = 7$ as from the origin.
   (c) The distance of $P$ from (1, 2) is twice its distance from $(-1, -2)$.
   (d) $P$ is on the normal to $x = 2t^2 + 1$, $y = 4t - 3$ at the point with parameter $t_1$.
   (e) If $L$ is (3, 5) and $M$ is (9, 2) then $P$ is such that $PL + PM = 8$.
   (f) If $A$ is a point on $x = 0$ and $B$ is a point on $y = 0$ where $AB = 6$ units, then $P$ is the mid-point of $AB$.
7. Find the locus of
   (a) the point which is equidistant from (1, 5) and (6, 8).
   (b) the point whose distance from (6, 2) is 3 times its distance from $(4, -5)$.
   (c) the point whose distance from (6, 5) is five units.
   (d) the mid-point of $AB$ where $AB$ is of length 6 units, $A$ lies on $x = 0$ and $B$ lies on $y = 0$.
   (e) the mid-point of $AB$ where $A$ is $(2, -1)$ and $B$ is any point on $3x^2 + 2y^2 = 6$.
   (f) the centroid of triangle $ABC$ where $A$ is (1, 1), $B$ is $(-1, -1)$ and $C$ is on the line $3x + 4y = 5$.
   (g) the mid-point of $PQ$ where $P$ is any point on $x = at^2$, $y = 2at$ and $Q$ is where the normal at $P$ meets the $x$-axis.

## 10.4 The Circle

**Exercise A**
1. Find the equations of the following circles:
   (a) With centre (0, 0) and radius 4.
   (b) With centre (6, 2) and radius 5.
   (c) With centre $(\frac{1}{2}, \frac{1}{4})$ and radius $2\sqrt{3}$.
   (d) With centre $(0.2, -1.4)$ and radius 3.72.
   (e) Such that (3, 4) and $(-1, 2)$ are the ends of a diameter.
2. Expand $(x-a)^2 + (y-b)^2 = r^2$ to give the form $x^2 + y^2 + 2gx + 2fy + c = 0$ and hence write $a$, $b$, and $c$ in terms of $g$, $f$, and $c$.
3. Some of the following curves are circles. Find which they are and give their centres and radii.
   (a) $x^2 + y^2 + 2x + 2y + 1 = 0$    (b) $x^2 + 2y^2 - x + y + 1 = 0$
   (c) $2x^2 + 2y^2 + x = 0$    (d) $x^2 + 3x + 4y - 6 = 0$
   (e) $3x^2 + 3y^2 + 6x + 12y = 8$    (f) $(x+2)^2 + y^2 = 6$
   (g) $x^2 + y^2 + 2xy + y + 1 = 0$    (h) $x^2 + y^2 + 2x + 2y + 10 = 0$
4. Find the equations of the circle:
   (a) with centre (0, 0) passing through $(-5, 12)$.
   (b) with centre $(-2, 4)$ passing through $(3, -1)$.
   (c) whose diameter is $AB$ where $A$ is (4, 2) and $B$ is (6, 8).

(d) which touches $x = 0$ at $(0, 3)$ and $y = 0$ at $(3, 0)$.
  (e) passing through $(3, 2)$, $(4, -5)$, and $(7, 9)$.
  (f) whose radius is the same as that of $x^2 + y^2 + x + y = 2$ and whose centre is halfway along the chord cut off from $y + x = 1$ by the given circle.
  (g) with centre $(4, -1)$ and touching the $y$-axis.
  (h) centred on $x = 6$ and passing through $(3, 4)$ and $(3, 9)$.
5. Show that the tangent at $(x_1, y_1)$ on $x^2 + y^2 + 2gx + 2fy + c = 0$ is
$$xx_1 + yy_1 + g(x + x_1) + f(y + y_1) + c = 0.$$
6. Find tangents to the following circles at the given points on the circles:
  (a) $x^2 + y^2 + 2x - y + 1 = 0$ at $(-1, 0)$
  (b) $x^2 + y^2 = 8$ at $(\sqrt{2}, \sqrt{6})$
  (c) $2x^2 + 2y^2 - 6x + y - 17 = 0$ at $(1, 3)$
  (d) $(x - 3)^2 + (y - 2)^2 = 1$ at $(2, 2)$
7. Find the length of the tangents to the following circles from the given points:
  (a) $x^2 + y^2 = 12$, $(4, 0)$  (b) $(x - 1)^2 + (y - 1)^2 = 2$, $(4, 5)$
  (c) $x^2 + y^2 + x = 6$, $(4, 8)$  (d) $3x^2 + 3y^2 + 2y + 3x = 0$, $(6, 6)$
  (e) $x^2 + y^2 + 4x - 2y + 7 = 0$, $(12, 5)$
8. Find a circle passing through the points of intersection of $x^2 + y^2 + 7x = 0$ and $x^2 + y^2 + 7y = 6$ which also passes through $(7, 9)$.
9. Find the two circles of radius 4 that pass through the points of intersection of $x^2 + y^2 = 9$ and $(x - 2)^2 + (y - 1)^2 = 4$.
10. Show that the common chord of $x^2 + y^2 + 2x - 2y - 3 = 0$ and $x^2 + y^2 + x + y - 1 = 0$ is $x - 3y = 2$.
11. Find the circle whose diameter is the common chord of $(x - 1)^2 + y^2 + 2y - 1 = 0$ and $x^2 + y^2 - 4y = 5$. Find the radius and centre.
12. Find which of the following pairs of circles are orthogonal:
  (a) $x^2 + y^2 + 2x - 4y - 5 = 0$ and $x^2 + y^2 - 2x + 6y - 9 = 0$
  (b) $x^2 + y^2 - 2x - 3y + 7 = 0$ and $x^2 + y^2 - x + 2y + 5 = 0$
  (c) $(x - 4)^2 + (y - 3)^2 = 16$ and $x^2 + y^2 = 9$
  (d) $x^2 + y^2 + 2x + y = 0$ and $2x^2 + 2y^2 + x + 2y + 2 = 0$
  (e) $3x^2 + 3y^2 + 2y + 5x + 3 = 0$ and $2x^2 + 2y^2 + 3x + 6y + 4 = 0$

**Exercise B**
1. Find the equations of the following circles:
  (a) centre $(0, 0)$ through $(8, -15)$;
  (b) passing through $(1, 1)$, $(2, 4)$, and $(7, 8)$;
  (c) of radius 3, whose centre is on $x = y$ and passing through $(4, 5)$;
  (d) whose centre is on $x = 0$ and having $y + x = 2$ and $x - y = 4$ as tangents;
  (e) whose centre is on $y = 2$ and having $y = 2x + 1$ and $4x - 2y - 1 = 0$ as tangents.
2. Find tangents to the following circles at the given points on them:
  (a) $(x + 4)^2 + (y - 1)^2 = 10$ at $(-1, 2)$
  (b) $3x^2 + 3y^2 + 2x + 4y = 7$ at $(-2, -1)$
3. Find the tangents to the following circles from the given points *not* on the circles:

71

(a) $x^2 + y^2 = 9$, (5, 0)  (b) $(x-2)^2 + (y-2)^2 = 4$, (4, 0)
(c) $x^2 + y^2 + 2x + 2y - 7 = 0$, (8, -1)

4. The line $y = 2x + 1$ cuts $x^2 + y^2 = 5$ at $A$ and $B$. Find where the tangents at $A$ and $B$ intersect.

5. Show that $y = 2x + 1$ cuts $x^2 + y^2 = 9$ in two real points. Find the range of values $c$ for which $y = 2x + c$ cuts the same circle at no real points and find the value of $c$ which makes the line a tangent.

6. (a) Show that the condition for $y = mx + c$ to be a tangent to $x^2 + y^2 = a^2$ is that $c = \pm a\sqrt{1 + m^2}$.
   (b) Find the tangents to $x^2 + y^2 = 9$ of gradient $\frac{1}{4}$.
   (c) Find the tangents to $2x^2 + 2y^2 - 12 = 0$ of gradient 0.2.

7. Find the lengths of the tangents to the given circles from the given points:
   (a) $x^2 + y^2 + 3x - 2y + 1 = 0$ from (2, 8)
   (b) $3x^2 + 3y^2 - x - 4 = 0$ from $(-6, -2)$
   (c) $(x-3)^2 + (y+2)^2 = 12$ from (4, 7)

8. Find the circle through the intersection points of $x^2 + y^2 + 4x + 6y - 3 = 0$ and $2x^2 + 2y^2 - 4x - 4y - 5 = 0$ which
   (a) passes through (0, 8).
   (b) passes through the centre of the first circle.
   (c) has $y = 1$ as a tangent.
   (d) cuts the $y$-axis at $90°$.

9. Find the common chord of the circles in question 10.

10. If $x^2 + y^2 - x + y = 6$ and $x^2 + y^2 + 2gx + 2y + c = 0$ have a common chord $3x + 4y = 2$, find $g$ and $c$. Find the centre and radius of the second circle.

11. Show that if two circles are orthogonal then the radii through their points of intersection are perpendicular to each other.

12. Show that if two circles have centres which are a distance '$d$' apart and have radii of $r_1$ and $r_2$ then $r_1^2 + r_2^2 = d^2$ is the condition that they are orthogonal. Hence establish which of the following pairs of circles are orthogonal:
    (a) $(x-6)^2 + y^2 = 25$ and $x^2 + (y-7)^2 = 60$
    (b) $x^2 + y^2 - 2x - 4y - 4 = 0$ and $x^2 + y^2 + 4x + 4y - 8 = 0$
    (c) $(x-2)^2 + (y-1)^2 = 16$ and $(x-3)^2 + (y-4)^2 = 25$
    (d) $x^2 + y^2 + 2x = 5$ and $x^2 + y^2 + 3x + y = 7$

## 10.5 The Parabola

**Exercise A**

1. Find which of the following are parabolas. Find in each case the vertex, focus and directrix.
   (a) $y^2 = 4x$       (b) $y^2 = 8x$       (c) $(y-1)^2 = 4x^2$
   (d) $x^2 = 8y$       (e) $(y-2)^2 = x$    (f) $y^2 + 2y = 4x$
   (g) $y^2 + x^2 = 2x$ (h) $4y^2 + y + 1 = 4(x-1)$  (i) $ay^2 = bx$

2. Find the Cartesian equations of the following parabolas:
   (a) vertex (0, 0), focus (2, 0)
   (b) vertex (1, 0), directrix $x = -3$

(c) focus (3, 0), directrix $x = 1$
(d) focus (1, 1), tangent at vertex $x = -1$
(e) vertex (2, 3), directrix $x = 1$
(f) vertex (1, 3), directrix $y = -1$

3. Show that $(at^2, 2at)$ lies on a parabola for all values of $t$.
4. Derive equations for the tangent and normal to the parabola $x = at^2$, $y = 2at$ at the point with parameter '$p$'.
5. If $O$ is the origin, $P$ is the point $(ap^2, 2ap)$ and $Q$ is the point where the normal at $P$ meets the x-axis, find $p$ if the area of $OPQ$ is $6a^2$ units.
6. Find the point $Q$ where the tangent at $P$ $(ap^2, 2ap)$ cuts the x-axis and hence show that if $S$ is the focus, $SQ = SP$. Hence show that angle $SQP$ = angle $SPQ$. How can we deduce from this fact that rays parallel to the axis of a parabolic mirror are always reflected towards the focus?
7. Show that tangents from points A, B with parameters $t_1$ and $t_2$ meet at $(at_1 t_2, a(t_1 + t_2))$.
8. Show that if $AB$ is a focal chord with parameters $t_1$ and $t_2$ then $t_1 t_2 = -1$.
9. Show from questions 7 and 8 that the tangents at the ends of a focal chord meet on the directrix.
10. Find the locus of the mid-points of parallel chords of a parabola $y^2 = 4x$ if they all have a gradient $m$.
11. Show that if $P$ and $Q$ lie on the parabola $y^2 = 4ax$ and $O$ is the origin where angle $POQ$ is $90°$, then the locus of the mid-points of $PQ$ is $y^2 = 2ax - 8a^2$. Show that this is a parabola and find its focus.
12. If $P$ is $(ap^2, 2ap)$ and $Q$ is $(aq^2, 2aq)$ and $PQ$ is a focal chord show that $OP$ cuts the directrix at $R(-a, 2aq)$. Hence find the area of triangle $PQR$.
13. Find the tangents to the parabola $y^2 = 4x$ from $(2, 3)$.
14. Find the tangent to $y^2 = 10x$ which is parallel to $2x + 3y = 5$.
15. Find the locus of mid-points of all chords of $y^2 = 8x$ which are parallel to $y = 3x$.

## Exercise B

1. Find which of the following curves are parabolas. State the vertex, directrix and focus in each case.
   (a) $y^2 = 2x$  (b) $x = 2t^2$, $y = 4t$  (c) $x = t^2$, $y = t$
   (d) $x = 2t$, $y = t^2$  (e) $4y = -(x-1)^2$  (f) $x^2 + y^2 = 4a(x+y)$
   (g) $y^2 + 8y = x + 1$  (h) $x^2 = 4a(x+y)$
2. Find the Cartesian equations of the following parabolas:
   (a) vertex (0, 0), focus (4, 0)  (b) directrix $x = -1$, focus (3, 0)
   (c) vertex (1, 2), focus (2, 2)  (d) vertex $(-1, 2)$, directrix $x = -3$
   (e) focus $(-2, 3)$, where the x-axis is the tangent at the vertex.
3. Find the length of the *latus rectum* of $y^2 = 4ax$.
4. Find the tangent and normal to $y^2 = 4x$ at $P$, (4, 4). If the normal meets the curve again at $Q$ find the co-ordinates of $Q$.
5. If tangents at $P$ and $Q$ to the parabola $x = at^2$, $y = 2at$ meet on the line $y = 4$ find the locus of the mid-point of $PQ$.
6. Show that if tangents to the parabola $y^2 = 4ax$ at $P$ and $Q$ meet at $(h, k)$ then $y = k$ bisects $PQ$.
7. Find, from the definition, the equation of the parabola whose directrix is $y - x = 4$ and focus is (2, 3).

73

8. Find the tangents from $(-2, 0)$ to $y^2 = 12x$.
9. Find where $y^2 = 4x$ cuts $x^2 = 4y$ and the angle between the tangents at the point of intersection.
10. $P$ is the point $(at^2, 2at)$ on $y^2 = 4ax$. The tangent at $P$ cuts the x-axis at $T$ and the normal at $P$ cuts the x-axis at $G$. $M$ is the point where the line parallel to the x-axis through $P$ cuts the directrix. If $S$ is the focus show that
    (a) $MPGS$ is a parallelogram and find its area
    (b) $MPST$ is a rhombus and find its area.

## 10.6 The Ellipse and Hyperbola

1. From the definition plot the ellipse with eccentricity $\frac{1}{2}$, one focus at $(3, 0)$ and corresponding directrix at $x = 12$. Find the second focus and directrix.
2. Plot the ellipse $4x^2 + y^2 = 4$.
3. Obtain the equation of the ellipse $\dfrac{x^2}{a^2} + \dfrac{y^2}{b^2} = 1$ from first principles.
4. Find the eccentricity, focus, and directrix of
    (a) $9x^2 + 16y^2 = 144$    (b) $9x^2 + 4y^2 = 16$
5. Find the equations of the following ellipses:
    (a) eccentricity $\frac{1}{4}$, focus at $(1, 0)$, centre the origin.
    (b) major axis of length 3, minor axis of length 2, centre at the origin.
    (c) focus at $(2, 0)$ and corresponding directrix $x = 8$, centre at $(0, 0)$.
6. Show that the sum of the distances from any point on the ellipse to the foci is constant and equal to $2a$.
7. Find the tangents and normals to the following ellipses at the given points:
    (a) $9x^2 + 4y^2 = 72$ at $(-2, 3)$
    (b) $x = 2\cos\theta$, $y = \sin\theta$ at $\theta = \frac{1}{3}\pi$
    (c) $5x^2 + 7y^2 = 34$ at $\left(\dfrac{1}{\sqrt{2}}, -\dfrac{3}{\sqrt{2}}\right)$
    (d) $b^2x^2 + 2a^2y^2 = 22a^2b^2$ at $(2a, 3b)$
8. Find $c$ if $y + 3x = c$ is a tangent to $3x^2 + y^2 = 6$.
9. Find the tangents of gradient 3 to the ellipse $5x^2 + 6y^2 = 30$ and find the normals at the points of contact.
10. If the tangent at $P$ cuts the x-axis at $T$ and the ordinate at $P$ cuts the x-axis at $N$, show that $ON \cdot OT = a^2$, where $O$ is the origin. If the normal at $P$ meets the x-axis at $M$ show that $OT \cdot NM = b^2$.
11. If a tangent at $P$ meets a directrix at $R$ and if $S$ is the focus nearest to the directrix show that angle $RSP = 90°$.
12. Tangents at $P$ and $Q$ meet at $R$. If $PQ$ cuts a directrix at $T$ show that angle $TSR = 90°$ where $S$ is the focus nearest the directrix on which $T$ lies.
13. From the definition sketch the hyperbola with eccentricity 2, with focus at $(4, 0)$ and directrix $x = 1$. Find the second focus and directrix.
14. Sketch $4x^2 - y^2 = 4$.

15. Establish the equation of the hyperbola, $\dfrac{x^2}{a^2} - \dfrac{y^2}{b^2} = 1$, from first principles.
16. Find the eccentricity, foci, and directrices of the following hyperbolas:
    (a) $\dfrac{x^2}{9} - \dfrac{y^2}{16} = 1$  (b) $3x^2 - 2y^2 = 6$
    (c) $x = 4 \sec \theta,\ y = 7 \tan \theta$.
17. Find the asymptotes of
    (a) $x^2 - 5y^2 = 1$   (b) $5x^2 - 4y^2 = 20$
    (c) $x = 2 \cosh u,\ y = 3 \sinh u$
18. Find the Cartesian and parametric equations of the following hyperbolas centred on (0, 0):
    (a) eccentricity 3, focus at (6, 0).
    (b) focus at (4, 0), corresponding directrix $x = 1$.
    (c) eccentricity 2 and cutting the axis at (3, 0).
    (d) with asymptotes $y = \pm 3x$, and focus (2, 0).
    (e) with asymptotes $y = \pm 8x$ and cutting $y = 0$ at (3, 0).
19. Find the tangents and normals to
    (a) $x^2 - 4y^2 = 9$, at $(2\sqrt{3}, -\tfrac{1}{2}\sqrt{3})$
    (b) $x = 2 \sec \theta,\ y = 5 \tan \theta$ at $\theta = \tfrac{2}{3}\pi$
    (c) $x = 5 \cosh u,\ y = 3 \sinh u$ at $u = 0$
    (d) $9x^2 - 4y^2 = 6$ at $(\sqrt{2}, \sqrt{3})$.
20. If $y = 2x + c$ is a tangent to $x^2 - 4y^2 = 9$, find $c$.
21. Find the condition on $c$ that $y = x + c$ cuts $9x^2 - 16y^2 = 144$ at (a) 2, (b) no, (c) 1 real point. Hence find the tangents with gradient 1, their points of contact and the normals at the points of contact.
22. Find the equations of the tangents from (1, 2) to $4x^2 - y^2 = 9$.
23. Find the equations of tangents and normals to the following rectangular hyperbolas at the given points:
    (a) $xy = 8$ at (1, 8)     (b) $xy = 3$ at $(\tfrac{1}{2}\sqrt{3}, 2\sqrt{3})$
    (c) $x = 3t,\ y = \dfrac{3}{t}$ at $t = 5$   (d) $x = ct,\ y = \dfrac{c}{t}$ at $t = t_1$
    (e) $xy = c^2$, at $(x_1, y_1)$
24. If the normal at the point $P$ where $t = 3$ to $x = 6t,\ y = \dfrac{6}{t}$ meets the curve again at $Q$, find the value of $t$ at $Q$ and the mid-point of $PQ$.
25. Show that $y + mx = 2\sqrt{m}$ is a tangent to $xy = 1$ for all real $m$.
26. Find $c$ in terms of $m$ and $b$ if $y = mx + c$ is to be a tangent to $xy = b^2$. Show that $m$ must be negative.
27. Find the locus of mid-points of chords of $xy = 16$ which pass through (4, 0).
28. Find the locus of mid-points of chords of $xy = c^2$ parallel to $y = 2x$.
29. Find the locus of mid-points of chords to $xy = 9$ which subtend a right angle at the focus.

75

## 10.7 Miscellaneous

1. Find the area of a triangle joining (1, 2), (3, −1) and (6, 4).
2. Find the centroid of a triangle whose vertices are at (3, 4), (4, 5) and (2, −1).
3. Find the mirror image of (a) (1, 3) in $2x + 3y = 7$ (b) (−1, 2) in $4y = 7x − 1$
4. Find the equations of the bisectors of the angles between $2x + y = 4$ and $5x + 2y = 6$.
5. $A$ is (1, 1), $B$ is (2, 7) and $C$ is (5, 8). $AD$ bisects angle $BAC$ cutting $BC$ at $D$. Show that $\dfrac{BD}{DC} = \dfrac{BA}{AC}$.
6. If the common chord of circles $C_1$ centre $(x_1, y_1)$ and radius $r_1$, and $C_2$ centre $(x_2, y_2)$ radius $r_2$ passes through the origin show that $x_1^2 + y_1^2 - r_1^2 = x_2^2 + y_2^2 - r_2^2$.
7. Show that $y = mx - 2\sqrt{1 + m^2}$ is always a tangent to $x^2 + y^2 = 4$. Hence find the common tangents to this circle and $y = \frac{1}{4}x^2 - 1$.
8. Find the value of $a$ if $y^2 = 4x$ and $x^2 + (y-2)^2 = a^2$ have a common tangent with gradient $-1$.
9. If $P$ is $(ap^2, 2ap)$ and $Q$ is $(aq^2, 2aq)$ and $PQ$ cuts the x-axis at $R$ where $3PR = PQ$ show that $q = -2p$. If the tangents at $P$ and $Q$ meet at $T$ show that the locus of $T$ is $2y^2 = -ax$.
10. Find the locus of the mid-point of $P$ and $Q$ on the parabola $y^2 = 4ax$ if $P$ and $Q$ subtend a right angle at the origin.
11. Show that the area of the parallelogram formed by tangents at the ends of conjugate diameters of $b^2x^2 + a^2y^2 = a^2b^2$ is $4ab$.
12. Show that the locus of the points of intersection of perpendicular tangents to $\dfrac{x^2}{a^2} + \dfrac{y^2}{b^2} = 1$ is $x^2 + y^2 = a^2 + b^2$. (The **director circle**.) If a tangent to the ellipse meets the director circle at $R$ and $S$, show that $OR$ and $OS$ are conjugate diameters that have been produced to meet the circle.
13. If $S_1$ and $S_2$ are foci of an ellipse and $P$ is any point on the ellipse, show that the normal at $P$ bisects angle $S_1PS_2$.
14. Find where the tangents to $x = a\cos\theta$, $y = b\sin\theta$ from points with parameters $\theta_1$ and $\theta_2$ meet.
15. Show that the eccentric angles of points at the ends of conjugate diameters differ by $\frac{1}{2}\pi$.
16. Show that $ax^2 + 2hxy + by^2 = 0$ represents a pair of straight lines through the origin and that the acute angle between them is $\tan^{-1}\left|\dfrac{2\sqrt{h^2 - ab}}{a+b}\right|$.
17. The tangent to $b^2x^2 - a^2y^2 = a^2b^2$ at any point $P$ $(a\sec\theta, b\tan\theta)$ cuts the asymptotes at $L$ and $M$. Show that (a) $LP = MP$. (b) the area of triangle $LOM$ is $ab$.
18. Write down the Cartesian equations of the following curves when the origin is translated to the given points, the axes remaining parallel to their original directions.
    (a) $x^2 + y^2 = a^2$, (2, 3)   (b) $y^2 = 4ax$, (−1, 2)

(c) $\dfrac{x^2}{a^2} + \dfrac{y^2}{b^2} = 1$, $(0, -3)$  (d) $y = 2x + 3$, $(-3, -4)$

(e) $xy = 9$, $(-5, -2)$  (f) $x^2 + 2x = y - 1$, $(2, 1)$

(g) $x = t^3$, $y = t + 1$, $(3, -1)$  (h) $x = \dfrac{t+1}{t-1}$, $y = t^2 + 2$, $(4, 6)$

19. Find the centres of the following conics and state the names of the curves:
    (a) $(x - a)^2 + (y - b)^2 = 1$  (b) $2x^2 + 2y^2 - 6x - 5y + 4 = 0$
    (c) $3x^2 + y^2 - 12x - 16 = 0$  (d) $xy - x + y = 7$
    (e) $x^2 - 2y^2 + 6x - 8y = 2$  (f) $x^2 + y^2 = 3xy$

# 11

# Determination of Laws

## 11.1 Notes and Formulae

If we wish to check that a particular law connecting two variables is true we may plot a graph of the observed values. If this is a straight line the law connecting the two variables can be easily identified; if it is a curve this is not possible. If, however, it proves possible to rephrase the law so that a straight line is obtained then the law can be identified.

The most general result is of the form

$$f(x, y) = m \cdot g(x, y) + c, \quad \text{where } m \text{ and } c \text{ are constants.}$$

From the table of values of $x$ and $y$ we calculate $Y = f(x, y)$ and $X = g(x, y)$ for each pair. We are thus checking the law $Y = mX + c$.

If a straight line connects the plotted points we have evidence that the law is true. It is then possible to find the constants $m$ and $c$ from the graph.

The use of logarithms is a useful device for some curves.

*Examples*

If the values of $a$ and $b$ are not known for the following curves, then they can be found from the graphs.

(a) $y = ax + b$  Plot $y$ against $x$; $a$ is the slope and $b$ is the intercept on the $y$-axis.

(b) $y = ax^b$  Take logs giving $\log y = b \log x + \log a$.
   Plot $Y = \log y$ against $X = \log x$ the slope will be $b$, and the intercept will be $\log a$.

(c) $y = ab^x$  Take logs giving $\log y = x \log b + \log a$.
   Plot $Y = \log y$ against $X = x$.
   Slope will be $\log b$, intercept will be $\log a$.

(d) $\dfrac{b}{x} + \dfrac{1}{y} = \dfrac{1}{a}$    Plot $Y = \dfrac{1}{y}$ against $X = \dfrac{1}{x}$.

Slope will be $-b$, intercept will be $\dfrac{1}{a}$.

(e) $xy = b(x+y) + a$   Plot $Y = xy$ against $X = x+y$.
Slope is $b$, intercept is $a$.

## 11.2 Examples on Determination of Laws

**Exercise A**

1. The following tables give corresponding values of $x$ and $y$. Plot suitable graphs to show that the points lie on a straight line of the form $y = mx + c$ and find the values of $m$ and $c$.

   (a)
   | $x$ | $-1$ | 0 | 1 | 2 | 3 |
   |---|---|---|---|---|---|
   | $y$ | $-3.5$ | $-1$ | 1.5 | 4 | 6.5 |

   (b)
   | $x$ | 0.5 | 1 | 1.5 | 2 |
   |---|---|---|---|---|
   | $y$ | $-3.05$ | $-3.8$ | $-4.55$ | $-5.3$ |

2. By plotting $y$ against $x^2$ show that the following values of $x$ and $y$ satisfy $y = ax^2 + b$ and find $a$ and $b$ from the graph.

   | $x$ | $-2$ | $-1$ | 0 | 1 | 2 | 3 |
   |---|---|---|---|---|---|---|
   | $y$ | 7 | 1 | $-1$ | 1 | 7 | 17 |

3. By plotting $\dfrac{1}{y}$ against $\dfrac{1}{x}$ show that $x$ and $y$ in the table below obey a law of the form $\dfrac{1}{y} = \dfrac{1}{bx} + c$ and find the values of $b$ and $c$ from your graph.

   | $x$ | 0.5 | 0.75 | 1 | 1.25 | 1.5 |
   |---|---|---|---|---|---|
   | $y$ | 0.33 | 0.43 | 0.50 | 0.56 | 0.60 |

4. $p$ and $q$ are believed to obey a law of the form $p = mq^n$. Plot $\log_{10} p$ against $\log_{10} q$ and show that this is true. Find the values of $m$ and $n$.

   | $q$ | 2 | 3 | 4 | 5 | 6 | 7 |
   |---|---|---|---|---|---|---|
   | $p$ | 7.071 | 12.990 | 20.000 | 27.951 | 36.742 | 46.301 |

5. Plot suitable graphs from the following tables to establish the suggested form of the law connecting the values of $x$ and $y$ and find the constants ($a$, $b$, or both) in each case.
   (a) $y = ax^2 + b$

   | $x$ | 0.1 | 0.2 | 0.3 | 0.4 | 0.5 | 0.6 |
   |---|---|---|---|---|---|---|
   | $y$ | $-1.269$ | $-1.176$ | $-1.021$ | $-0.804$ | $-0.525$ | $-0.184$ |

   (b) $y = a(1+x)^b$

| $x$ | $-0.5$ | $-0.25$ | 0 | 0.25 | 0.5 | 0.75 |
|---|---|---|---|---|---|---|
| $y$ | 0.1682 | 0.1861 | 0.2000 | 0.2115 | 0.2213 | 0.2300 |

(c) $y = ab^x$

| $x$ | 0.25 | 0.375 | 0.5 | 0.625 | 0.75 |
|---|---|---|---|---|---|
| $y$ | 9.87 | 11.32 | 12.99 | 14.90 | 17.10 |

(d) $x + \dfrac{1}{a} = axy$

| $x$ | 5 | 10 | 15 | 20 | 25 |
|---|---|---|---|---|---|
| $y$ | 10.00 | 7.50 | 6.67 | 6.25 | 6.00 |

(e) $y^2 = axy + b$

| $x$ | $-2.56$ | $-1.63$ | $-0.49$ | 1.4 | 2.7 | 3.6 |
|---|---|---|---|---|---|---|
| $y$ | 2.904 | 2.105 | 1.275 | 0.521 | 0.330 | 0.259 |

(f) $y = a^{1+x}$

| $x$ | 1 | 3 | 5 | 7 | 9 | 11 |
|---|---|---|---|---|---|---|
| $y$ | 12.25 | 150.1 | 1838 | 22520 | 275900 | 3379000 |

# Exercise B

1. Plot $\sqrt{y}$ against $x$ to show that the following table gives a relationship between $x$ and $y$ in the form $\sqrt{y} = kx - l$. Find $k$ and $l$.

| $x$ | 0.5 | 0.7 | 0.9 | 1.1 | 1.3 | 1.5 |
|---|---|---|---|---|---|---|
| $y$ | 0.25 | 2.25 | 6.25 | 12.25 | 20.25 | 30.25 |

2. Plot $x^2 y$ against $\dfrac{x}{y}$ to verify a law of the form $x^2 y = a\dfrac{x}{y} + b$ between $x$ and $y$ from the following table and find values for $a$ and $b$.

| $x$ | 0.1 | 0.2 | 0.3 | 0.4 | 0.5 |
|---|---|---|---|---|---|
| $y$ | 499.9 | 124.9 | 55.4 | 30.4 | 19.69 |

3. $s$ and $t$ are thought to vary according to the law $s = kl^{-t}$. Plot a graph of $\log_{10} s$ against $t$ to establish this and find values for $k$ and $l$.

| $t$ | 5 | 10 | 15 | 20 | 25 | 30 |
|---|---|---|---|---|---|---|
| $s$ | $5.76 \times 10^{-3}$ | $2.37 \times 10^{-5}$ | $9.76 \times 10^{-8}$ | $4.02 \times 10^{-10}$ | $1.65 \times 10^{-12}$ | $6.80 \times 10^{-15}$ |

4. Plot suitable graphs from the tables below to establish the suggested form of law and find the constants $k$ and $l$ in each case.

(a) $y = \dfrac{k}{x^2} + l$

| $x$ | 0.8 | 1.5 | 1.9 | 2.0 | 2.6 |
|---|---|---|---|---|---|
| $y$ | 12.99 | 7.84 | 7.07 | 6.95 | 6.48 |

(b) $y = kx^l$

| x | 1 | 3 | 5 | 7 | 9 | 11 |
|---|---|---|---|---|---|---|
| y | 4.90 | 33.51 | 81.92 | 147.61 | 229.15 | 325.56 |

(c) $ky = l^{x+2}$

| x | -2 | -1 | 0 | 1 | 2 | 3 |
|---|---|---|---|---|---|---|
| y | 4 | 1.2 | 0.36 | 0.108 | 0.0324 | 0.00972 |

(d) $\dfrac{1}{y} = \dfrac{k}{x^3} + \dfrac{1}{l}$

| x | 0.1 | 0.12 | 0.14 | 0.16 | 0.18 | 0.20 |
|---|---|---|---|---|---|---|
| y | 0.00048 | 0.00082 | 0.00131 | 0.00195 | 0.00277 | 0.00380 |

(e) $l(x+y) = \dfrac{ky}{x^2}$

| x | 10 | 15 | 20 | 25 | 30 | 35 | 40 |
|---|---|---|---|---|---|---|---|
| y | -10.067 | -15.045 | -20.033 | -25.027 | -30.022 | -35.019 | -40.017 |

# 12

# Complex Numbers

## 12.1 Notes and Formulae

Complex numbers can all be written in the form $a + ib$ where $a$ and $b$ are real numbers and i stands for $\sqrt{-1}$. (Electrical engineers use the letter j in place of i). Hence $i^2 = -1$, $i^3 = -i$, and $i^4 = 1$.
$a - ib$ is called the conjugate of $a + ib$
$(a + ib)(a - ib) = a^2 + b^2$.
If $x + iy = a + ib$, then $x = a$ and $y = b$. This process is called "equating real and imaginary parts" and is a very useful tool. Thus it can be proved that

$$\cos 3\theta + i \sin 3\theta = (\cos \theta + i \sin \theta)^3 = 4\cos^3 \theta - 3\cos \theta + i(3 \sin \theta - 4 \sin^3 \theta)$$

Then, equating real and imaginary parts

$$\cos 3\theta = 4\cos^3 \theta - 3\cos \theta \quad \text{and} \quad \sin 3\theta = 3\sin \theta - 4\sin^3 \theta$$

Addition, subtraction, multiplication and division can be performed with complex numbers using the same rules as for real numbers. In division the denominator should be multiplied by its conjugate thus making the denominator a real number, e.g.

$$\frac{2+3i}{1-2i} = \frac{(2+3i)(1+2i)}{(1-2i)(1+2i)} = \frac{-5+7i}{5} = -1 + \frac{7}{5}i$$

In the Argand diagram the number $x + iy$ is plotted in its Cartesian co-ordinates $(x, y)$; thus the $x$-axis represents all real numbers and the $y$-axis all purely imaginary numbers.

All complex numbers may be represented in the modulus–argument, or polar, form corresponding to polar co-ordinates $(r, \theta)$ on the Argand diagram. If a complex number is $a + ib$, then
$a + ib = r(\cos\theta + i\sin\theta)$ where $r$ is the modulus and $\theta$ is the argument.
The modulus, $r$, written $|a + ib| = \sqrt{a^2 + b^2}$.
The argument is given by $\arg(a + ib) = \tan^{-1}\frac{b}{a}$, where $-\pi < \theta \leq \pi$
$(\cos\theta_1 + i\sin\theta_1)(\cos\theta_2 + i\sin\theta_2) = \cos(\theta_1 + \theta_2) + i\sin(\theta_1 + \theta_2)$
$$\frac{\cos\theta_1 + i\sin\theta_1}{\cos\theta_2 + i\sin\theta_2} = \cos(\theta_1 - \theta_2) + i\sin(\theta_1 - \theta_2)$$

## 12.2 Basic Operations

1. Evaluate in the form $a + ib$
   (a) $(2 + 3i) + (5 - 2i)$, $(3 - 4i) - (2 - 2i)$, $(-1 - \sqrt{3}i) + (2 + \sqrt{3}i)$
   (b) $(3 - 2i) + (3 + 2i)$, $(\sqrt{3} + i) - (2\sqrt{3} + 3i)$,
   $(-1 + \sqrt{2}i) - (1 - 2\sqrt{2}i)$
   (c) $(4 + 3i)(3 + 2i)$, $(4 + i)(2 - 2i)$, $(-1 - i)(4 + 2i)$
   (d) $(1 + \sqrt{3}i)(1 - \sqrt{3}i)$, $(3 + 4i)(2 - 7i)$, $(3 - 2i)(4 - 3i)$
   (e) $(3 + 4i)^2$, $(i - 2i)^2$, $(3 + i)^2$
   (f) $\dfrac{3+4i}{3-4i}$, $\dfrac{3-2i}{1+i}$, $\dfrac{5+12i}{3+2i}$
   (g) $\dfrac{(2+i)(3+i)}{1-i}$, $\dfrac{(4-3i)(1+i)}{1+2i}$, $(1+i)^3$

2. Express the following in the form $a + ib$:
   (a) $\sqrt{2}\left(\cos\dfrac{\pi}{4} + i\sin\dfrac{\pi}{4}\right)$  (b) $5(\cos 0 + i\sin 0)$
   (c) $6\left(\cos\dfrac{\pi}{3} + i\sin\dfrac{\pi}{3}\right)$  (d) $4(\cos\pi + i\sin\pi)$
   (e) $2\left(\cos\dfrac{2\pi}{3} + i\sin\dfrac{2\pi}{3}\right)$  (f) $2\sqrt{2}\left(\cos\dfrac{7\pi}{4} + i\sin\dfrac{7\pi}{4}\right)$
   (g) $3(\cos 2\pi + i\sin 2\pi)$  (h) $10\left(\cos\dfrac{3\pi}{4} + i\sin\dfrac{3\pi}{4}\right)$
   (i) $7(\cos 3\pi + i\sin 3\pi)$  (j) $8\left(\cos\dfrac{10\pi}{3} + i\sin\dfrac{10\pi}{3}\right)$

3. Express the following in polar form, $r(\cos\theta + i\sin\theta)$, writing your answers in the form $(r, \theta)$:

(a) $\frac{1}{2}\sqrt{3} + \frac{1}{2}i$   (b) $\frac{1}{\sqrt{2}} + \frac{1}{\sqrt{2}}i$   (c) $-\frac{1}{2} + \frac{1}{2}\sqrt{3}i$

(d) $-\frac{1}{2}\sqrt{3} - \frac{1}{2}i$   (e) $-1 - i$   (f) $\sqrt{3} - i$

(g) $3 + 4i$   (h) $-6i$   (i) $12 + 5i$

4. Write down the results of the following multiplications (i) in $(r, \theta)$ form and (ii) in $a + ib$ form.

(a) $2\left(\cos\frac{\pi}{6} + i\sin\frac{\pi}{6}\right) 3\left(\cos\frac{\pi}{3} + i\sin\frac{\pi}{3}\right)$

(b) $4\left(\cos\frac{\pi}{4} + i\sin\frac{\pi}{4}\right) 2\left(\cos\frac{3\pi}{4} + i\sin\frac{3\pi}{4}\right)$

(c) $6\left(\cos\frac{\pi}{3} + i\sin\frac{\pi}{3}\right) 4\left(\cos\frac{\pi}{2} + i\sin\frac{\pi}{2}\right)$

(d) $2\sqrt{2}\left(\cos\frac{\pi}{4} + i\sin\frac{\pi}{4}\right) \sqrt{2}\left(\cos\frac{\pi}{2} + i\sin\frac{\pi}{2}\right)$

(e) $\left[5\left(\cos\frac{5\pi}{3} + i\sin\frac{5\pi}{3}\right)\right]^2$

5. Write down the results of the following divisions (i) in $(r, \theta)$ form. (ii) in $a + ib$ form.

(a) $6\left(\cos\frac{\pi}{2} + i\sin\frac{\pi}{2}\right) / 2\left(\cos\frac{\pi}{4} + i\sin\frac{\pi}{4}\right)$

(b) $8\left(\cos\frac{3\pi}{4} + i\sin\frac{3\pi}{4}\right) / 4\left(\cos\frac{\pi}{4} + i\sin\frac{\pi}{4}\right)$

(c) $5\left(\cos\frac{4\pi}{3} + i\sin\frac{4\pi}{3}\right) / 2\left(\cos\frac{\pi}{3} + i\sin\frac{\pi}{3}\right)$

(d) $4\left(\cos\frac{\pi}{2} + i\sin\frac{\pi}{2}\right) / \left(\cos\frac{\pi}{3} + i\sin\frac{\pi}{3}\right)$

(e) $2\left(\cos 2\pi + i\sin 2\pi\right) / \frac{1}{2}\left(\cos\frac{4\pi}{3} + i\sin\frac{4\pi}{3}\right)$

## 12.3 Complex Roots

**Exercise A**

1. Solve the equations
   (a) $z^2 - 2z + 5 = 0$   (b) $z^2 + 2z + 3 = 0$
   (c) $z^2 - z + 2 = 0$   (d) $2z^2 + z + 4 = 0$
2. Write down the conjugates of $1 + 2i$, $3 - 2i$, $3i$, $-4 - 5i$, $\cos\theta + i\sin\theta$.
3. Factorise $z^3 - 1 = 0$ and hence find all three cube roots of unity. Square

each of the complex roots and compare with the other complex root. What do you find?
4. Each of the following equations has a simple real root which can be found by inspection. Use this root to factorise the left hand side and hence solve the equations completely.
    (a) $z^3 + z - 2 = 0$    (b) $z^3 - z^2 - 4 = 0$
    (c) $z^3 - 2z^2 - 9 = 0$    (d) $2z^3 - 3z^2 + 3z - 1 = 0$
5. Find the four roots of $z^4 - 16 = 0$.

**Exercise B**
1. Solve the equations
    (a) $z^2 - 4z + 5 = 0$    (b) $z^2 + z + 3 = 0$
    (c) $z^2 - 3z + 3 = 0$    (d) $3z^2 - 3z + 1 = 0$
2. Write down the conjugates of $2 + i$, $2 - 3i$, $-2 + 5i$, $4$, $\cos \phi - i \sin \phi$.
3. Factorise $z^3 + 1 = 0$ and hence find the three cube roots of $-1$. What numbers do the squares of these roots represent?
4. Each of the following equations has a simple real root which can be found by inspection. Use this root to factorise the left hand side and hence solve the equations completely.
    (a) $z^3 - z^2 + 2 = 0$    (b) $z^3 + z^2 + 4 = 0$
    (c) $z^3 - 10z - 24 = 0$    (d) $2z^3 + 3z^2 + 2z - 2 = 0$
5. Find the two real roots of the equation $z^6 - 1 = 0$, and find two complex roots for $z^2$ from the quadratic factor. (After dealing with section 12.5 you should then be able to find all six roots of the equation.)

## 12.4 Geometrical Representation

To save space we shall often refer to "the point $3 + 4i$" instead of using the more precise statement "the point on the Argand diagram which represents $3 + 4i$.
1. Plot the following sets of points on an Argand diagram, and name the shape of the figure so formed:
    (a) $i, 2 + 3i, 5i, -2 + 3i$
    (b) $1 + i, 3 + 3i, 4i$
    (c) $1 + i, 4 + 2i, 2 + 3i, -1 + 2i$
    (d) $-1 - i, 2, \frac{1}{2} + 4\frac{1}{2}i, -2\frac{1}{2} + 3\frac{1}{2}i$
2. Factorise $x^4 - 1$ completely and hence find the four fourth roots of unity. Plot these on an Argand diagram and name the figure so formed.
3. Plot the points $2 + 15i$, $10 + 9i$ and $4i$ and *prove* that the triangle is isosceles.
4. Plot the three cube roots of unity on an Argand diagram and *prove* that they are the vertices of an equilateral triangle.
5. Plot the points $-1 - i$, $3 + 2i$, and $-6 + 14i$ and *prove* that they form a right-angled triangle.
6. On the same Argand diagram plot $z = 3 + 4i$, $\bar{z}$ (the conjugate of $z$), $-z$ and $-\bar{z}$. What shape is formed?
7. If $z_1 = 5 + 3i$ and $z_2 = 2 + i$, plot these points on an Argand diagram and also plot (a) $z_1 + z_2$ (b) $z_1 - z_2$ (c) $z_2 - z_1$ (d) $2z_2$ (e) $z_1 + 2z_2$ (f) $\bar{z}_1 + \bar{z}_2$

83

8. On an Argand diagram plot the points $z = 5 - 2i$, $\bar{z}$ and $-z$. Write down the complex numbers representing the mid points of the sides of the triangle.
9. If $z_1 = 4 + 3i$ and $z_2 = 2 + 2i$ plot these points on an Argand diagram and also plot (a) $z_1 + z_2$ (b) $z_1 - z_2$ (c) $z_2 - z_1$ (d) $3z_2$ (e) $z_1 + 3z_2$ (f) $\bar{z}_1 + \bar{z}_2$
10. Show that the points $3i$, $9 + 6i$ and $5 - 2i$ lie on a circle and find the centre of that circle.

## 12.5 Equating Real and Imaginary Parts

1. Find the values of $x$ and $y$ in the following equations:
   (a) $2x + 3iy = 0$ 
   (b) $(x - 1) + i(y + 2) = 0$
   (c) $3x + y + i(x + y - 2) = 0$ 
   (d) $(2x + y + 2) + i(3x + 2y + 5) = 0$
   (e) $2x + 2iy = 1 - y - 3ix$
2. If $z = x + iy$ and $\bar{z} = x - iy$, solve the following:
   (a) $2z + \bar{z} = 3 - 2i$ 
   (b) $3z - 2\bar{z} = -1 + 15i$
   (c) $(z + 5) - 2(\bar{z} - 3) = 1 - 2i$
3. If $z = x + iy$, find in terms of $x$ and $y$
   (a) $z + \bar{z}$ (b) $z^2$ (c) $\dfrac{1}{z} + \dfrac{1}{\bar{z}}$ (d) $z\bar{z}$ (e) $\dfrac{1}{z^2}$
4. By assuming $\sqrt{5 + 12i} = a + ib$ and squaring both sides, find the two values of the square root.
5. By the same method find
   (a) $\sqrt{-3 + 4i}$ (b) $\sqrt{-21 - 20i}$ (c) $\sqrt{15 - 8i}$ (d) $\sqrt{-2 - 2\sqrt{3}i}$
6. Find the values of $x$ and $y$ if
   (a) $(x + iy)(3 + i) = 11 - 3i$ 
   (b) $(2x + iy)(1 + i) = -2 + 6i$
   (c) $\dfrac{x + iy}{2 - i} = 5 + 3i$ 
   (d) $\dfrac{2x - 5iy}{3 - 2i} = 4 + i$
   (e) $\dfrac{2x + iy}{4 - i} = 3 + 2i$ 
   (f) $\dfrac{5x + 7iy}{5 - 2i} = 2 + 5i$

## 12.6 Miscellaneous

1. If $z = \cos\theta + i\sin\theta$ show that $z^{-1} = \cos\theta - i\sin\theta$. (Multiply numerator and denominator by the conjugate.) Hence show that
$$z + z^{-1} = 2\cos\theta \quad \text{and} \quad z - z^{-1} = 2i\sin\theta$$
2. If $z = \cos\theta + i\sin\theta$ show that $z^2 = \cos 2\theta + i\sin 2\theta$. Hence show that
$$z^2 + z^{-2} = 2\cos 2\theta \quad \text{and} \quad z^2 - z^{-2} = 2i\sin 2\theta$$
Also deduce that $z^4 = \cos 4\theta + i\sin 4\theta$ and hence find expressions for $\cos 4\theta$ and $\sin 4\theta$ in terms of $\cos\theta$ and $\sin\theta$.
3. If $\omega$ and $\omega^2$ are the complex cube roots of unity show that
   (a) $\omega^4 = \omega$ 
   (b) $\omega + \omega^2 = -1$ 
   (c) $\omega + \omega^2 + \omega^3 = 0$
   (d) $\omega^3 + \omega^4 = -\omega^5$ 
   (e) $\sqrt{\omega} = \omega^2$
4. If $P$ is the point $z = 4 + i$ on an Argand diagram, what are the complex

numbers represented by the following points? (a) the reflection of $P$ in the $x$-axis, (b) the reflection in the $y$-axis, (c) in the origin, (d) in the line $y = x$, (e) in the line $x + y = 0$.

5. (a) $\alpha$, $\beta$ are the roots of $2z^2 - 2z + 1 = 0$. Express $\alpha$ and $\beta$ in modulus–argument form. With these values for $\alpha$ and $\beta$ simplify the expression $\dfrac{\alpha\beta + 1 + 5i}{\alpha + \beta - 1 + i}$.

(b) $z_1$ and $z_2$ are the roots of $z^2 - 6z + 13 = 0$. Show on an Argand diagram the points $z_1$, $z_2$, $z_1 + z_2$, $z_1 z_2$, $z_1/z_2$ and $z_2/z_1$.

6. Simplify $\dfrac{(1+i)^4}{(1-i)^2}$ expressing the result in polar form.

7. If the roots of the equation $(1 - i)z^2 - z + 2 - i = 0$ are $\alpha$ and $\beta$, find and simplify the value of $\alpha + \beta$ and $\alpha\beta$. Find the equation whose roots are $\alpha + 2\beta$ and $2\alpha + \beta$.

8. If $\omega$ is one of the complex cube roots of unity, show that $\omega^2$ is the other and that $1 + \omega + \omega^2 = 0$. Hence simplify $(1 + 2\omega + \omega^2)^2$ and $(1 + \omega + 2\omega^2)^2$ and also find the sum and product of these two expressions.

9. $z_1 = \dfrac{a}{1-i}$, $z_2 = \dfrac{b}{2+i}$ and $2z_1 + z_2 = 3$.

Find $a$ and $b$. With these values for $a$ and $b$ find the complex number representing the mid-point of the line joining the points representing $z_1$ and $z_2$ on an Argand diagram.

10. Given that $z = 3 + 4i$ find $\dfrac{1}{z}$ and the two square roots of $z$ in the form $a + ib$. Plot $z$, $-(z + \bar{z} - 1)$, $\bar{z}$, and the two square roots of $z$ on an Argand diagram. What is the shape of the figure formed by these points?

11. If $z = 2 + i$ is a root of the equation $z^3 + az^2 + bz + c = 0$ where the coefficients are all real, find two equations involving $a$, $b$, and $c$. If $c = 10$ find $a$ and $b$.

12. Show that $z = 1 + i$ and $z = -1 - i$ are both roots of $z^4 + 4 = 0$ and hence write down all four roots.

# 13

# Differentiation

## 13.1 Notes and Formulae

**Standard Results**

Differential coefficients marked with an asterisk are included so as to make the table more or less complete. Students are not normally expected to learn them.

| $y$ | $\dfrac{dy}{dx}$ | $y$ | $\dfrac{dy}{dx}$ |
|---|---|---|---|
| $x^n$ | $nx^{n-1}$ | $e^x$ | $e^x$ |
| $\sin x$ | $\cos x$ | *$a^x$ | $a^x \log_e a$ |
| $\cos x$ | $-\sin x$ | $\log_e x$ | $\dfrac{1}{x}$ |
| $\tan x$ | $\sec^2 x$ | *$\log_a x$ | $\dfrac{1}{x}\log_a e$ |
| *$\csc x$ | $-\csc x \cot x$ | $\sin^{-1}\dfrac{x}{a}$ | $\dfrac{1}{\sqrt{a^2 - x^2}}$ |
| *$\sec x$ | $\sec x \tan x$ | $\cos^{-1}\dfrac{x}{a}$ | $-\dfrac{1}{\sqrt{a^2 - x^2}}$ |
| *$\cot x$ | $-\csc^2 x$ | $\tan^{-1}\dfrac{x}{a}$ | $\dfrac{a}{a^2 + x^2}$ |

### Differentiation Methods

*Function of a Function*

If $y = f(u)$, where $u = \phi(x)$ then $\dfrac{dy}{dx} = \dfrac{dy}{du}\dfrac{du}{dx}$

*Product*

If $y = uv$ (where $u$ and $v$ are both functions of $x$)

$$\dfrac{dy}{dx} = v\dfrac{du}{dx} + u\dfrac{dv}{dx}$$

*Quotient*

If $y = \dfrac{u}{v}$ (where $u$ and $v$ are both functions of $x$)

$$\dfrac{dy}{dx} = \dfrac{v\dfrac{du}{dx} - u\dfrac{dv}{dx}}{v^2}$$

### Implicit Differentiation

*Example*

If $x^2 - y^2 + 2x - 4y + 10 = 0$

on differentiating both sides with respect to $x$  $\quad 2x - 2y\dfrac{dy}{dx} + 2 - 4\dfrac{dy}{dx} = 0$

whence $\qquad (-2y - 4)\dfrac{dy}{dx} = -2x - 2$

and $\qquad \dfrac{dy}{dx} = \dfrac{x+1}{y+2}$

## Parametric Forms

*Example*

If $x = 2a \sin \theta$ and $y = a \cos 2\theta$

$$\frac{dx}{d\theta} = 2a \cos \theta \quad \text{and} \quad \frac{dy}{d\theta} = -2a \sin 2\theta$$

$$\frac{dy}{dx} = \frac{dy}{d\theta}\frac{d\theta}{dx} = \frac{dy}{d\theta}\bigg/\frac{dx}{d\theta} = -\frac{2a \sin 2\theta}{2a \cos \theta} = -2 \sin \theta$$

## Logarithmic differentiation

*Example*

Given $y = \dfrac{(x-1)\sqrt{x^2-1}}{x+1}$ to find $\dfrac{dy}{dx}$.

Taking logs: $\log y = \log(x-1) + \tfrac{1}{2}\log(x^2-1) - \log(x+1)$

differentiating, $\dfrac{1}{y}\dfrac{dy}{dx} = \dfrac{1}{x-1} + \dfrac{x}{x^2-1} - \dfrac{1}{x+1}$

$$= \frac{x+1+x-x+1}{x^2-1} = \frac{x+2}{x^2-1}$$

$$\frac{dy}{dx} = \frac{(x-1)\sqrt{x^2-1}(x+2)}{(x+1)(x^2-1)} = \frac{(x+2)\sqrt{x^2-1}}{(x+1)^2}$$

## 13.2 Differentiation of Powers of x

No attempt has been made to vary the independent variable – all differentiations are with respect to $x$.

1. If $\delta y$ is the small increase in $y$ resulting from a small increase $\delta x$ in $x$, calculate $\dfrac{\delta y}{\delta x}$ and hence find $\dfrac{dy}{dx}$ for the following functions of $x$:

   (a) $y = 5x+2$    (b) $y = 2x^2$    (c) $y = 3x^2+x$
   (d) $y = x^3$    (e) $y = x^2 - 2x + 2$    (f) $y = x^3 + x^2$

   In questions 2 to 5 differentiate the functions given with respect to $x$, using the standard rule.

2. (a) $x^4$    (b) $x^6$    (c) $7x^3$
   (d) $-6x$    (e) $\tfrac{1}{4}x^4$    (f) $3x^2 - 6x + 2$
   (g) $4x^3 + x - 5$    (h) $x^3 - 2x^2 + 3x - 4$

3. (a) $x^{-2}$    (b) $x^{-4}$    (c) $\dfrac{1}{x}$    (d) $\dfrac{1}{x^3}$
   (e) $\dfrac{4}{x^2}$    (f) $\dfrac{2}{x^4}+1$    (g) $\dfrac{2}{x^2}-\dfrac{1}{x}$    (h) $\dfrac{1}{x^3}+\dfrac{1}{x}+x^2$

4. (a) $x^{1/2}$    (b) $\sqrt[3]{x}$    (c) $x^{2/3}$    (d) $x^{1\frac{1}{2}}$
   (e) $(\sqrt[3]{x})^2$    (f) $\dfrac{1}{\sqrt[3]{x}}$    (g) $\dfrac{1}{\sqrt[4]{x}}$    (h) $\dfrac{1}{\sqrt{x}}+\dfrac{1}{\sqrt[3]{x}}$

5. (a) $x(x^2+2)$  (b) $(x^2+2)^2$  (c) $x^2(x^2+x+1)$

(d) $\dfrac{x^3+1}{x^2}$  (e) $\dfrac{x^2+2x+3}{x}$  (f) $x\left(x^3+\dfrac{1}{x^3}\right)$

(g) $\left(x+\dfrac{1}{x}\right)^2$  (h) $(x+2)(x^2-3)$  (i) $(x^2+1)(\sqrt{x}+1)$

## 13.3 Function of a Function

Further practice in this principle is given in section 13.4.

1. Find $\dfrac{dy}{dx}$ if  (a) $y=u^3$, where $u=2x+3$

(b) $y=2u^5$ where $u=x^2+1$  (c) $y=\dfrac{1}{u^2}$ where $u=3x+1$

(d) $y=z+\dfrac{1}{z}$ where $z=x^3-1$  (e) $y=2z^2-\dfrac{1}{z^2}$, where $z=x^2-4$

2. Find $\dfrac{dy}{dx}$ if  (a) $y=2u^2$ where $u=x^2+2$

(b) $y=3u^4$ where $u=2x+5$  (c) $y=\dfrac{1}{u^3}$ where $u=2x^2-3$

(d) $y=z^2+\dfrac{1}{z^2}$ and $z=x^2+6$  (e) $y=u^2\left(u-\dfrac{1}{u}\right)$ and $u=x^3+x$

3. Differentiate with respect to $x$

(a) $(x^3+5)^4$  (b) $(x^3+x^2+1)^5$  (c) $\dfrac{1}{x^4+4}$

(d) $\sqrt[3]{2x^2-1}$  (e) $\dfrac{1}{\sqrt{2x^2-1}}$  (f) $\sqrt[4]{x^2-16}$

(g) $\dfrac{1}{(2x+1)^4}$  (h) $\dfrac{1}{(3x-1)\sqrt[3]{3x-1}}$

4. Differentiate with respect to $x$

(a) $(x^2+2)^3$  (b) $(x^2+2x+3)^4$  (c) $\dfrac{1}{x^3-8}$

(d) $\sqrt{x^2+4}$  (e) $\dfrac{1}{\sqrt{x^2+4}}$  (f) $\sqrt[3]{x^3+8}$

(g) $\dfrac{1}{(2x^2+3)^3}$  (h) $\dfrac{1}{(x^2+6)\sqrt{x^2+6}}$

## 13.4 Further Basic Differentiation

Differentiate with respect to $x$ (or $t$ where appropriate) all the functions given in the following questions:

1. (a) $\sin 2x$ (b) $\cos 3x$ (c) $\tan \tfrac{1}{2}x$
   (d) $\sin(2x+1)$ (e) $\cos(3x+2)$ (f) $\tan(\tfrac{1}{2}x+\tfrac{1}{4}\pi)$
   (g) $\sin(2x+\tfrac{1}{6}\pi)$ (h) $\tan(3x+\tfrac{1}{2}\pi)$ (i) $\cos(nx+\pi)$
   (j) $2\tan(\tfrac{1}{2}x+2)$ (k) $6\cos(\tfrac{1}{3}x+\tfrac{1}{4}\pi)$ (l) $4\sin(\tfrac{1}{4}x-\tfrac{1}{2}\pi)$

2. (a) $\sin^2 x$ (b) $\cos^3 x$ (c) $\sqrt{\tan x}$
   (d) $\operatorname{cosec} x$ (e) $\cot x$ (f) $\tan^2 x$
   (g) $\cos^4 x$ (h) $\operatorname{cosec}^2 x$ (i) $\cot^2 x$
   (j) $\sin^2(2x+1)$ (k) $\cos^2(x^2+1)$ (l) $\tan^3(x^3+1)$
   (m) $\sqrt{\sin(2x+\tfrac{1}{6}\pi)}$ (n) $\dfrac{1}{\sqrt{\cos 2x}}$ (o) $\dfrac{1}{\tan^2 4x}$

3. (a) $(1+\sin x)^3$ (b) $\sin^2 x + \cos^2 x$ (c) $(\sin x + \cos x)^2$
   (d) $\cos^2 x - \sin^2 x$ (e) $\tan x + \cot x$ (f) $\sec x$

4. (a) $e^{2x}$ (b) $e^{2x+1}$ (c) $e^{x^2}$ (d) $e^{-2x}$
   (e) $e^{1/x}$ (f) $4e^{x/2}$ (g) $e^{x^2+1}$ (h) $e^{-x^2}$
   (i) $e^{\sin x}$ (j) $e^{\tan x}$ (k) $e^{\sin 2x}$ (l) $e^{\tan \tfrac{1}{2}x}$

5. (a) $\log 2x$ (b) $\log(x+1)$ (c) $\log x^2$
   (d) $\log(x^2+1)$ (e) $\log 2x^2$ (f) $\log(x^2+2x+3)$
   (g) $\log \sin x$ (h) $\log \sin 2x$ (i) $\log \tan x$
   (j) $\log \cos(x^2+1)$ (k) $\log(e^{x+1})$ (l) $\log e^{\sin x}$

6. (a) $\sin^{-1} x$ (b) $\tan^{-1} x$ (c) $\cos^{-1}\tfrac{1}{2}x$ (d) $\tan^{-1}\tfrac{1}{2}x$
   (e) $\sin^{-1}(2x-1)$ (f) $\tan^{-1}\tfrac{1}{4}x$ (g) $\sin^{-1}\tfrac{2}{3}x$

## 13.5 Products and Quotients

**Exercise A**

Differentiate the expressions given in the following questions:

1. (a) $x^2 \sin x$ (b) $xe^x$ (c) $x \log x$ (d) $e^x \cos x$
   (e) $\sin x \cos x$ (f) $x^2(1+\sin x)$ (g) $x \sin 2x$ (h) $x\sqrt{1+x^2}$

2. (a) $\dfrac{x^2}{\sin x}$ (b) $\dfrac{e^x}{x}$ (c) $\dfrac{\log x}{x}$ (d) $\dfrac{\sin x}{e^x}$
   (e) $\dfrac{e^x}{\sin x}$ (f) $\dfrac{x^2}{1+\sin x}$ (g) $\dfrac{x}{\sin 2x}$ (h) $\dfrac{2x}{\sqrt{1+x^2}}$

3. (a) $e^x \tan x$ (b) $\dfrac{e^x}{\tan x}$ (c) $e^x \log x$ (d) $\dfrac{e^x}{\log x}$
   (e) $\sin x \log x$ (f) $\dfrac{\log x}{\sin x}$ (g) $x \sin^{-1} x$ (h) $\dfrac{\sin^{-1} x}{x}$

4. (a) $\dfrac{x \cos x}{1+x^2}$ (b) $\dfrac{x^2 e^x}{\sin x}$ (c) $\dfrac{x \log x}{\cos x}$
   (d) $\dfrac{(1+x^2)\sin x}{e^x}$ (e) $\dfrac{e^x \tan x}{x^2-5}$ (f) $\dfrac{\sec x \tan x}{e^x}$

**Exercise B**

1. (a) $x^3 \cos x$ (b) $xe^{-x}$ (c) $x^2 \log x$ (d) $e^{-x} \sin x$
   (e) $\sin x \tan x$ (f) $x^3(1-\cos x)$ (g) $x^2 \cos 2x$ (h) $x^2 \sqrt{1+x}$

89

2. (a) $\dfrac{x}{\cos x}$ (b) $\dfrac{x}{e^x}$ (c) $\dfrac{x^2}{\log x}$ (d) $\dfrac{\cos x}{e^x}$

(e) $\dfrac{e^x}{\tan x}$ (f) $\dfrac{x^3}{1-\cos x}$ (g) $\dfrac{x^2}{\cos 2x}$ (h) $\dfrac{x^2}{\sqrt{1-x^2}}$

3. (a) $\tan x \log x$ (b) $\dfrac{\log x}{\tan x}$ (c) $e^{-x}\tan x$ (d) $\dfrac{\tan x}{e^{-x}}$

(e) $\sqrt{1+x^2}\, e^x$ (f) $\dfrac{e^x}{\sqrt{1+x^2}}$ (g) $x^2 \tan^{-1} x$ (h) $\dfrac{\tan^{-1} x}{x^2}$

4. (a) $\dfrac{x^2 \tan x}{1+x}$ (b) $\dfrac{xe^{-x}}{\cos x}$ (c) $\dfrac{x^2 \log x}{\sin x}$

(d) $\dfrac{\cos x \cos 3x}{x^3}$ (e) $\dfrac{\sin x \cos 2x}{\sin 3x}$ (f) $\dfrac{(1+x^2)\tan^{-1} x}{x^3}$

## 13.6 Further Differentiation Methods

1. Find the value of $\dfrac{dy}{dx}$ in the following cases by "implicit differentiation".

(a) $y^2 = 4ax$ (b) $\dfrac{x^2}{16} + \dfrac{y^2}{9} = 1$ (c) $xy = 12$

(d) $x^2 + y^2 + 2x + 4y - 20 = 0$ (e) $x^2 + xy + y^2 = 10$

(f) $y^3 = 4(x-1)$ (g) $x^2 y^2 - 4x = 0$

(h) $\sqrt{x} + \sqrt{y} = 1$ (i) $x^3 + y^3 = 3xy$

2. Find $\dfrac{dy}{dx}$ from the following pairs of equations, giving your answer in terms of the parameter $\theta$ or $t$.

(a) $x = a\cos\theta,\ y = b\sin\theta$ (b) $x = ct,\ y = \dfrac{c}{t}$

(c) $x = \dfrac{\sin\theta}{1+\cos\theta},\ y = \dfrac{\cos\theta}{1+\sin\theta}$ (d) $x = t + \dfrac{1}{t},\ y = t - \dfrac{1}{t}$

(e) $x = t^3 + 1,\ y = 2t^2$ (f) $x = \dfrac{3t}{t-3},\ y = \dfrac{2t}{t-2}$

(g) $x = e^{\sin\theta},\ y = e^{\cos\theta}$ (h) $x = \log\sin\theta,\ y = \log\cos\theta$

3. Use logarithmic differentiation to find $\dfrac{dy}{dx}$ where $y$ is equal to the following functions of $x$:

(a) $\sqrt{\dfrac{1-x}{1+x}}$ (b) $\dfrac{(x^2+1)^3}{\sqrt{2x-1}}$ (c) $x^{\sin x}$ (d) $x^{\tan x}$

(e) $\dfrac{x^4 \sin^4 x}{\cos^5 x}$ (f) $x^x$ (g) $x^{e^x}$ (h) $\dfrac{(x+1)^5 \cos^3 x}{\sin^6 x}$

(i) $100^x$

## 13.7 Miscellaneous

1. Differentiate the following with respect to $x$:
   (a) $\dfrac{3-x}{1-3x}$    (b) $\sin^2 4x$    (c) $e^{\sin 2x}$
   (d) $\log \dfrac{x-1}{x+1}$    (e) $(x^2+1)\tan 2x$    (f) $\sqrt{1+9x^2}$
   (g) $\sin^2 x \cos^2 x$    (h) $e^{\sin \pi x}$    (i) $\sin(x^2+1)$
   (j) $\dfrac{x^2+1}{\sqrt{x^2-1}}$    (k) $\dfrac{\sin(x+\frac{\pi}{6})}{\cos(x-\frac{\pi}{2})}$    (l) $\log\sqrt{\dfrac{1-x}{1+x}}$

2. Differentiate with respect to $x$
   (a) $x \log x$    (b) $\operatorname{cosec}^2 2x$
   (c) $10^{2x}$    (d) $\dfrac{1}{x^3}\sqrt{1+x^2}$
   (e) $\log\{x+\sqrt{x^2+1}\}$    (f) $e^{\pi x}\cos \pi x$
   (g) $\log\left(\dfrac{x^3+1}{x^2+1}\right)$    (h) $\cot^2 2x$;
   (i) $(1-2x)^2\sqrt{1-2x}$    (j) $\log(e^{2x}-1)$
   (k) $x \sin x \sin 2x$    (l) $\log_{10} x$
   (m) $\left(\dfrac{1}{x}-\dfrac{1}{x^2}\right)^2$    (n) $\sin^{-1}\tfrac{1}{2}(x-1)$
   (o) $(e^x \cos x)^3$    (p) $\log\{x+\sqrt{x^2-a^2}\}$
   (q) $\log \tan \tfrac{1}{2}x$    (r) $\cos^{-1}\dfrac{x}{a}$
   (s) $\log \sin^2 2x$

3. Differentiate $\log \dfrac{(2x-1)^4 (x+1)^2}{\sqrt{x^2-1}}$

4. Use logarithmic differentiation to find $\dfrac{dy}{dx}$ if
   $y = \dfrac{(1+x^2)e^{-x}}{1-x^2}$ and hence find $\dfrac{dy}{dx}$ when $x = -2$.

5. Find $\dfrac{dy}{dx}$ for the curve $(x+y-1)(x-y+2)+12 = 0$ and hence find the gradients of the tangents where $x = 0$.

6. For each of the following curves find $\dfrac{dy}{dx}$ and $\dfrac{d^2y}{dx^2}$ in terms of $x$ and $y$:
   (a) $\dfrac{x^2}{a^2} + \dfrac{y^2}{b^2} = 1$
   (b) $x^2 + y^2 + 2x + 2y - 2 = 0$
   (c) $x^2 + y^2 = 4xy$

7. If $y = e^x \cos \pi x$ find the equation of the tangent at $x = 1$ and also prove that $\dfrac{d^2y}{dx^2} - 2\dfrac{dy}{dx} + (\pi^2 + 1)y = 0$.

8. If $y = e^{ax} \sin bx$ show that $\dfrac{d^2y}{dx^2} - 2a\dfrac{dy}{dx} + (a^2 + b^2)y = 0$.

9. Express $y = \dfrac{4x - 5}{(x-1)(2x-3)}$ in partial fractions and hence find $\dfrac{dy}{dx}$ and $\dfrac{d^2y}{dx^2}$ when $x = 2$.

10. Differentiate *from first principles*    (a) $\dfrac{1}{x}$    (b) $\sqrt{x}$    (c) $\dfrac{1}{\sqrt{x}}$

11. Prove that $\dfrac{d}{dx}\sin^{-1}x = \dfrac{1}{\sqrt{1-x^2}}$ and hence find the tangent to the curve $\sin^{-1}y = \sin^{-1}x + \tfrac{1}{2}\pi$ at the point where $x = \tfrac{1}{2}$.

12. The curve called the Cycloid has the parametric equations $x = \theta - \sin\theta$, $y = 1 - \cos\theta$.
    Show that $\dfrac{dy}{dx} = \cot\tfrac{1}{2}\theta$ and that $\dfrac{d^2y}{dx^2} + \dfrac{1}{y^2} = 0$. Find also the equation of the tangent to the curve at $\theta = \tfrac{1}{2}\pi$.

13. The Astroid is given by the equations $x = a\cos^3\theta$, $y = a\sin^3\theta$. Find the Cartesian equation of the curve and the tangent at the point $\theta = \alpha$.

14. The Cardioid is given by $x = a(2\cos\theta + \cos 2\theta)$, $y = a(2\sin\theta + \sin 2\theta)$. Find $\dfrac{dy}{dx}$ in terms of $\theta$ in its simplest form.

# 14

# Applications of Differentiation

## 14.1 Notes and Formulae

### Gradients, Tangents and Normals

The value of $\dfrac{dy}{dx}$ at any point $(x_1, y_1)$ on a curve gives the gradient of the curve (i.e. of the tangent) at this point. Call this '$m$'.

    Equation of tangent is $y - y_1 = m(x - x_1)$.

    Equation of normal is $y - y_1 = -\dfrac{1}{m}(x - x_1)$.

## Maxima, Minima, and Points of Inflexion

For a maximum $\dfrac{dy}{dx} = 0$ and $\dfrac{d^2y}{dx^2} < 0$

For a minimum $\dfrac{dy}{dx} = 0$ and $\dfrac{d^2y}{dx^2} > 0$

For a point of inflexion $\dfrac{dy}{dx} =$ any value and $\dfrac{d^2y}{dx^2} = 0$.

There are a few exceptions to the simple test of using the sign of $\dfrac{d^2y}{dx^2}$ to distinguish between a maximum and a minimum. If in any doubt find the sign of $\dfrac{dy}{dx}$ immediately on each side of the turning point. If the signs reading from left to right (i.e. $x$ increasing) are $+, 0, -$ then the point is a maximum, but if they are $-, 0, +$ the point is a minimum. If the signs read $-, 0, -$ or $+, 0, +$ then there is a point of inflexion. This method is also useful if the second differential coefficient is difficult to calculate.

In problems the essential steps are to express in mathematical terms the quantity for which a maximum or minimum is required and then use the conditions given in the question to ensure that it contains only one variable.

## Small Increments and Connected Rates of Change

As $\lim\limits_{\delta x \to 0} \dfrac{\delta y}{\delta x} = \dfrac{dy}{dx}$ then $\delta y \approx \dfrac{dy}{dx} \cdot \delta x$ and this principle can be used to calculate approximately the change in one variable caused by a small change in another.

Calculations involving interconnected rates of change involve linking the appropriate differential coefficients, e.g. in the case of water flowing into a vessel we could have

$$\dfrac{dV}{dt} = \dfrac{dV}{dx} \dfrac{dx}{dt}$$

where $V =$ volume, $x =$ depth of water, $t =$ time, $\dfrac{dV}{dt} =$ rate of flow and $\dfrac{dx}{dt} =$ rate of change of depth.

## Distance, Velocity and Acceleration

If $s =$ distance, $v =$ velocity, $t =$ time then

$$v = \dfrac{ds}{dt} \text{ and acceleration} = \dfrac{dv}{dt} = \dfrac{d^2s}{dt^2} = v\dfrac{dv}{ds}$$

## 14.2 Gradients, Tangents and Normals

1. Find the gradients of the following curves at $x = 0$ and at $x = 2$:

93

(a) $y = 3x + 2$  (b) $y = x^2 + 3x$  (c) $y = 4x + 2 - x^2$
(d) $y = 2x - x^2$  (e) $y = \dfrac{1}{x^2 + 4}$  (f) $y = e^{2x}$
(g) $y = \log(x + 1)$  (h) $y = \log\sqrt{x^2 + 1}$

2. Find the gradients of the following curves at $x = 0$ and $x = \tfrac{1}{3}\pi$:
   (a) $y = \sin x$  (b) $y = \tan x$  (c) $y = \cos 2x$
   (d) $y = \sin(2x + \tfrac{1}{6}\pi)$  (e) $y = \cot(\tfrac{1}{2}x - \tfrac{1}{6}\pi)$

3. Find the equation of the tangent to each of the curves in question 1 (b) to (h) at the point where $x = 1$.

4. Find the equation of the normal to each of the curves in question 1 (b) to (h) when $x = 1$.

5. Find the points of intersection of the tangents to the following curves at the points given:
   (a) $y = 4x - x^2$  at $x = 1$ and $x = 3$
   (b) $y = x^3 - 8x^2 + 15x$  at $x = 0$ and $x = 3$
   (c) $y = 2x + 3 - x^2$  at $x = 0$ and $x = 2$
   (d) $y = 16 - x^4$  at $x = -2$ and $x = 1$

## 14.3 Maxima, Minima and Points of Inflexion

1. Find the turning points on the following curves, stating which are maxima and which are minima:
   (a) $y = 4x - x^2$  (b) $y = x^2 - 6x + 3$
   (c) $y = x^3 - 12x$  (d) $y = x^3 - 3x^2 - 9x + 1$
   (e) $y = 2x^3 + 3x^2 - 36x + 10$  (f) $y = x^4 - 8x^2$

2. Find the points of inflexion of the curves in question 1 (c), (d) and (e).

3. Find the first maximum and the first minimum occurring after $x = 0$ on the following curves:
   (a) $y = \sin(x - \tfrac{1}{3}\pi)$  (b) $y = \cos(3x + \tfrac{1}{2}\pi)$
   (c) $y = \tfrac{1}{2}\sin(\tfrac{1}{2}x + \tfrac{1}{6}\pi)$

4. Find the maximum and minimum points of the following:
   (a) $y = x + \dfrac{1}{x - 2}$  (b) $y = x^2 + \dfrac{16}{x}$
   (c) $y = \dfrac{x^2 + 4}{2x - 3}$  (d) $y = x - 2\sin x$  $(0 \leqslant x \leqslant 2\pi)$
   (e) $y = 4x - \tan x$  $(0 \leqslant x \leqslant \pi)$

5. Find the turning points on
   (a) $y = xe^x$  (b) $y = e^x \sin x$  $(0 \leqslant x \leqslant 2\pi)$  (c) $y = x \log x$

## 14.4 Maxima and Minima Problems

1. A rectangular area is to be enclosed in the corner of a field using the hedges of two sides of the field together with 60 metres of fencing to form the other two sides. Find the dimensions of the rectangle to give the maximum area.

2. A farmer forms a rectangular enclosure using the straight hedge of a field

for one side and 100 metres of fencing to form the other three sides. Find the dimensions to give maximum area.

3. An open tray is made by cutting squares of side $x$ cm from each corner of a rectangular piece of sheet metal 20 cm by 30 cm and folding up the sides. Find the value of $x$ to the nearest millimetre to give the tray with the maximum volume.

4. The cost of running a certain machine is $C$ pence per minute where $C = 120 + 15x(x-4)$, $x$ being the speed of the machine in revolutions per second. Find the speed which gives the minimum cost.

5. Find the location of the rectangle of maximum area which can be drawn in the first quadrant with two vertices on the curve $y = 4x - x^2$ and the other two on the $x$-axis.

6. The total cost of manufacturing $x$ articles is $C = a + bx + cx^2$. Find the value of $x$ which gives the minimum cost *per article*.

7. From a strip of paper $2l$ cm wide two circles are drawn, each touching one edge of the strip and touching each other (the line of the centres being perpendicular to the edges of the strip). Find the values of the radii so as to give the minimum area.

8. An open cylindrical tank of radius $r$ has a fixed volume $V$. Show that for minimum surface area $r^3 = \dfrac{V}{\pi}$.

9. A cone of thin metal plate of radius '$r$' and volume $V$ without a base is made by cutting a circle from a square of metal and then removing a sector from the circle. Show that the area of metal used (including the sector removed and the corners of the square) is

$$4\left(\dfrac{V^2 + \pi^2 r^6}{\pi^2 r^4}\right)$$

and that this is a minimum for fixed $V$ if $r^3 = \dfrac{\sqrt{2V}}{\pi}$.

10. $ABCD$ is a trapezium where $AB = CD = BC = 4$ m. $BC$ is parallel to $AD$ and $AD$ is greater than $BC$. Find the angle $BAD$ so that the area is a maximum.

11. A closed circular cylinder has a total surface area of $600\pi$ cm². Find the value of the radius which gives the maximum volume.

12. From a strip of width '$a$' cm a square and a circle are cut, together taking up the whole width of the strip. If the circle has a radius of '$x$' cm, show that the minimum total area of the square and the circle occurs when $x = \dfrac{2a}{\pi + 4}$ and that the area is then approximately $0.44a^2$.

## 14.5 Small Increments and Connected Rates of Change

All questions in this section should be answered by using the appropriate differential coefficients.

1. The radius of a circle is increased from '$r$' to '$r + \delta r$' cm. Find the approximate increase in area. What is the precise increase?
2. A spherical balloon of radius '$r$' cm is inflated so that the radius becomes '$r + \delta r$'. Find its approximate increase in volume.
3. A square of side 10 cm is enlarged so that the side is increased by 1 mm. Find the approximate increase in area.

4. The radius of an ink bolt (assumed circular) increases at the rate of 2 mm per second. If the bolt initially had a radius of 0.5 cm find the approximate increase in area between 5 and 5.5 seconds.
5. A balloon of radius 20 cm is inflated so that its radius is increased by 5 mm. Find the approximate increase in volume.
6. Liquid flows into a vessel so that the volume after '$t$' seconds is given by $V = 10 + 2t + t^2$ cm$^3$. Find the approximate increase in volume between 5 and 5.1 seconds.
7. Under certain conditions the volume (in cm$^3$) and the pressure (in g cm$^{-2}$) of a gas are connected by the relation $pv = 300$. What is the approximate change in volume if the pressure increases 15 to 15.3 g cm$^{-2}$.
8. The displacement of a piston from its mean position is given by $s = 10 \sin \omega t$ cm. If $\omega = 100$ find the change in displacement between $t = 0.01$ and $t = 0.011$ seconds.
9. A radioactive substance decays so that after '$t$' days its mass is given by $m = Ae^{-kt}$. Find the approximate loss of mass between day 100 and day 101 if $A = 1000$ g and $k = 0.01$.
10. A tank has the form of an inverted cone in which the diameter of the top is equal to its height. Water is poured in at the rate of 10 cm$^3$ per second. At what rate does the level of the water rise when the depth is 10 cm?
11. A stylus on a record player moves across a record of radius 15 cm at a speed of 3 mm per minute towards the centre of the record. At what rate is the area left to play decreasing after 10 minutes?
12. A wine glass is so shaped that the volume contained is $\frac{1}{2}\pi h^2$ where $h$ is the depth of liquid in the glass. If wine is poured in at a rate of 5 cm$^3$ per second at what rate is the depth increasing when the depth is 5 cm?

## 14.6 Distance, Velocity and Acceleration

All the following questions refer to motion in a straight line, where $t$ = time in seconds, $v$ = velocity in cm s$^{-1}$ and $a$ = acceleration in cm s$^{-2}$.
1. If $s = 40t + 5t^2$ find (a) the velocity, (b) the acceleration, when $t = 0$ and when $t = 5$.
2. Given that $s = 80t - 16t^2$ find (a) the velocity at $t = 0$, (b) the velocity at $t = 2$, (c) the velocity when $t = 5$ and interpret the result, (d) when the velocity is zero, (e) the acceleration, and (f) the maximum velocity. (In this question only, distances are in feet.)
3. Given that $s = 75t^2 - t^3$ find (a) the velocity at $t = 0$ and $t = 2$, (b) the time when the velocity is zero, (c) the acceleration when $t = 0$ and when $t = 2$ and (d) the time when the acceleration is zero.
4. One particle moves along a straight line so that $s = 50t - 2t^2$ and a second one moves so that $s = 10t + 2t^2$. Find (a) when and how far from their common starting point they meet again, (b) at what time their speeds are equal, and (c) whether their accelerations are ever equal.
5. If $s = 20t^2 - 4t^3$ find (a) the velocity when $t = 0$ and $t = 2$, (b) the time when $v = 0$, (c) the acceleration when $t = 0$ and $t = 2$ and (d) the time when the acceleration is zero.
6. Two particles move in a straight line so that their distances from a common reference point are given by $s = 35t - t^2$ and $s = 100 + 2t^2$ respectively. Find (a) how far apart they are at the start, (b) at what time

they meet, (c) at what speed they approach each other when $t = 5$ and (d) at what time their velocities are equal.
7. If $s = 5 + 3 \sin t + 4 \cos t$ where $s$ is measured from point 0, find (a) the distance from 0 when $t = 0$ and when $t = \frac{1}{2}\pi$, (b) an expression for the velocity, (c) an expression for the acceleration and (d) the relation connecting distance and acceleration.

## 14.7 Miscellaneous

1. Find the points where the tangents to the following curves at the points given meet each other:
   (a) $y^2 = 16x$          at (1, 4) and (4, 8)
   (b) $x^2 + y^2 = 25$     at (3, 4) and (4, $-3$)
   (c) $xy = 15$           at (3, 5) and (5, 3)
   (d) $x^2 + y^2 - 8x - 6y = 0$    at (4, 8) and (8, 0)
2. Find the turning points on the curve $y = \cos 2x - 2\cos x$   $(0 \leq x \leq 2\pi)$.
3. For the curve $x^2 + y^2 + xy - x + y - 2 = 0$ find $\frac{dy}{dx}$ in terms of $x$ and $y$. Find the gradient of the curve at each of the four points where it cuts the axes.
4. Find the maximum and minimum values and the point of inflexion of the curve $y = 4\sin^2 x + 3\cos^2 x$, where $0 \leq x \leq 2\pi$. Sketch the curve for this range.
5. Investigate the turning points for the curve $y = 2x^4 + px^2$. For what values of $p$ does this curve have three turning points? If $p = -1$ find the turning points and sketch the curve.
6. Find the turning points for the curve $y = x^2 + 1 + \frac{1}{x^2}$. Sketch the curves $y = x^2 + 1$ and $y = x^2 + 1 + \frac{1}{x^2}$ on the same diagram.
7. A piece of wire of fixed length '$a$' cm is bent to form an enclosed figure comprising a semi-circle mounted on a rectangle, the diameter which forms one side of the rectangle being omitted. If the radius of the semi-circle is $x$ cm show that the maximum area is obtained when $x = \frac{a}{4 + \pi}$.
8. For a gas at constant temperature the relation between the pressure and the volume is given by $pv = C$. Initially $p = 1500 \text{ gm cm}^{-2}$ and $v = 1000 \text{ cm}^3$. If the volume is reduced by 5 cm³ per second find the rate of change of pressure when the volume is 500 cm³.
9. An area comprises an equilateral triangle joined to a rectangle and sharing one side with the rectangle. If the perimeter of the pentagon so formed is '$a$' cm, and the side of the triangle is '$x$' cm, find the value of $x$ which gives the maximum area to 3 significant figures.
10. A vessel is such that the volume '$V$' cm³ is given by
$$V = \frac{\pi}{12}(y^3 + 30y^2 + 300y)$$ where '$y$' cm is the depth of liquid in the vessel.
Water is poured in at the rate of 40 cm³ per second. Find the rate at which the depth is increasing when $y = 6$.

# 15

# Integration and Differential Equations

## 15.1 Notes and Formulae

### Standard Integrals

Students are not normally expected to learn those integrals marked with an asterisk.

$$\int x^n \, dx = \frac{x^{n+1}}{n+1} + C \quad (n \neq -1)$$

\* $\int \sec x \, dx = \log(\tfrac{1}{2}x + \tfrac{1}{4}\pi) + C$
$\phantom{\int \sec x \, dx} = \log(\sec x + \tan x) + C$

$$\int \frac{1}{x} \, dx = \log_e x + C$$

\* $\int \csc x \, dx = \log \tan \tfrac{1}{2}x + C$

$$\int \sin x \, dx = -\cos x + C$$

$$\int e^x \, dx = e^x + C$$

$$\int \cos x \, dx = \sin x + C$$

$$\int \frac{1}{\sqrt{1-x^2}} \, dx = \sin^{-1} x + C$$

$$\int \sec^2 x \, dx = \tan x + C$$

$$\int \frac{1}{1+x^2} \, dx = \tan^{-1} x + C$$

$$\int \csc^2 x \, dx = -\cot x + C$$

$$\int \frac{1}{\sqrt{a^2-x^2}} \, dx = \sin^{-1} \frac{x}{a} + C$$

$$\int \tan x \, dx = \log \sec x + C$$

$$\int \frac{1}{a^2+x^2} \, dx = \frac{1}{a} \tan^{-1} \frac{x}{a}$$

$$\int \cot x \, dx = \log \sin x + C$$

### Methods of Integration

$$\int f'(u) \frac{du}{dx} \, dx = f(u) + C$$

It is important to be able to recognise this type as the result of differentiating a function of a function.

e.g. $\int \sin^2 x \cos x \, dx = \frac{1}{3} \sin^3 x + C$

*Integration by Substitution*

If the variable is changed, the 'dx' and also the limits (if any) must be changed. A useful substitution is

$t = \tan \frac{1}{2}\theta$; then $d\theta = \dfrac{2\,dt}{1+t^2}$, $\sin\theta = \dfrac{2t}{1+t^2}$, $\cos\theta = \dfrac{1-t^2}{1+t^2}$, $\tan\theta = \dfrac{2t}{1-t^2}$.

*Example 1* $\quad \displaystyle\int \sqrt{16-x^2}\,dx$

Let $x = 4\sin\theta$; $dx = 4\cos\theta\,d\theta$; $\sqrt{16-x^2} = 4\cos\theta$.

$$\text{Integral} = \int 4\cos\theta\,4\cos\theta\,d\theta$$
$$= 8\int 2\cos^2\theta\,d\theta$$
$$= 8\int (\cos 2\theta + 1)\,d\theta$$
$$= 4\sin 2\theta + 8\theta + C$$
$$= 8\sin\theta\cos\theta + 8\theta + C$$
$$= \tfrac{1}{2}x\sqrt{16-x^2} + 8\sin^{-1}\tfrac{1}{4}x + C$$

*Example 2*

$$\int_0^{\frac{1}{2}\pi} \frac{1}{1+2\cos\theta}\,d\theta$$

Put $t = \tan\frac{1}{2}\theta$; $\quad dt = \dfrac{2\,dt}{1+t^2}$.

When $\theta = 0$, $t = 0$.
When $\theta = \frac{1}{2}\pi$, $t = 1$.

$$\text{Integral} = \int_0^1 \frac{2\,dt}{(1+t^2)\left(1 + \dfrac{2-2t^2}{1+t^2}\right)}$$
$$= \int_0^1 \frac{2\,dt}{3-t^2} = \int_0^1 \frac{2\,dt}{(\sqrt{3}-t)(\sqrt{3}+t)}$$
$$= \frac{1}{2\sqrt{3}} \int_0^1 \frac{1}{\sqrt{3}-t} + \frac{1}{\sqrt{3}+t}\,dt$$
$$= \left[\frac{1}{2\sqrt{3}} \log \frac{\sqrt{3}+t}{\sqrt{3}-t}\right]_0^1$$
$$= \frac{1}{2\sqrt{3}} \log \frac{\sqrt{3}+1}{\sqrt{3}-1}$$

*Integration by Parts*

$$\int u\, dv = uv - \int v\, dv \text{ (This is the simplest form to remember.)}$$

*Example*

$$\int x \sec^2 x\, dx = x \tan x - \int \tan x\, dx$$

$$= x \tan x - \log \sec x + C$$

*Differential Equations (by separation of variables)*

If an equation involving $x$, $y$ and $\dfrac{dy}{dx}$ can be put in the form $f(y)\dfrac{dy}{dx} = g(x)$ then it can be solved by integrating both sides with respect to $x$.

$$\int f(y)\frac{dy}{dx}dx = \int g(x)\,dx$$

Changing the variable gives

$$\int f(y)\,dy = \int g(x)\,dx$$

For example,
$$\frac{dy}{dx} = 3xy$$

rearrange to
$$\frac{1}{y}\frac{dy}{dx} = 3x$$

this then gives
$$\int \frac{1}{y}\,dy = \int 3x\,dx$$

$$\log y = \tfrac{3}{2}x^2 + c$$

## 15.2 Integration of Simple Functions

1. Integrate the following powers of $x$:

   (a) $x^5$   (b) $x^{-3}$   (c) $3x^3$   (d) $\dfrac{2}{x^2}$

   (e) $\tfrac{1}{2}x^{1/2}$   (f) $\sqrt{x}$   (g) $x^{-2/3}$   (h) $\dfrac{1}{x\sqrt{x}}$

2. Integrate the following functions:

   (a) $x^2 + 3x + 5$   (b) $x^4 + 1 + \dfrac{2}{x^2} + \dfrac{2}{x^3}$   (c) $\sqrt{x} + \dfrac{1}{\sqrt{x}}$

   (d) $(x-1)(x-2)$   (e) $(x^2 - 3)^2$   (f) $\left(x + \dfrac{1}{x}\right)^2$

   (g) $\left(x^2 + \dfrac{1}{x}\right)^2$

3. Evaluate

(a) $\int_0^3 5x^4 \, dx$  (b) $\int_1^4 \frac{1}{x^2} \, dx$

(c) $\int_2^3 (4x^2 + 3x + 2) \, dx$  (d) $\int_{-1}^2 (3x+1)(x+3) \, dx$

(e) $\int_{-3}^{-1} (2x+3)^2 \, dx$

4. Evaluate

(a) $\int_{-1}^1 3x^2 - \frac{1}{x^2} \, dx$  (b) $\int_1^3 \frac{(x^2+1)^2}{x^2} \, dx$

(c) $\int_1^2 \frac{(2x+3)(3x+2)}{x^4} \, dx$  (d) $\int_1^3 \left(\sqrt{x} + \frac{1}{\sqrt{x}}\right)^2 dx$

(e) $\int_1^{27} \sqrt[3]{x} + \frac{1}{\sqrt[3]{x}} \, dx$

5. (a) $\int \cos 3x \, dx$  (b) $\int \sin 2x \, dx$  (c) $\int \sin(x + \tfrac{1}{6}\pi) \, dx$

(d) $\int \cos(x - \tfrac{1}{3}\pi) \, dx$  (e) $\int \sec^2 4x \, dx$  (f) $\int \sec^2(2x - \tfrac{1}{6}\pi) \, dx$

6. (a) $\int_0^{\pi/4} \sin 4x \, dx$  (b) $\int_{-\pi/6}^{\pi/6} \cos 2x \, dx$

(c) $\int_0^{\pi/3} \sec^2(x - \tfrac{1}{6}\pi) \, dx$  (d) $\int_0^{\pi/3} \sin(\tfrac{1}{2}x + \tfrac{1}{6}\pi) \, dx$

(e) $\int_{\pi/6}^{\pi/3} \cos(3x - \tfrac{1}{2}\pi) \, dx$  (f) $\int_{\pi/6}^{\pi/3} \operatorname{cosec}^2 x \, dx$

7. (a) $\int 6e^t \, dt$  (b) $\int e^{t-1} \, dt$  (c) $\int_0^1 e^{2x} \, dx$  (d) $\int_{-1}^2 e^{-x} \, dx$

8. (a) $\int \frac{1}{2x} \, dx$  (b) $\int \frac{1}{x+1} \, dx$  (c) $\int_1^2 \frac{1}{2x+1} \, dx$  (d) $\int_3^4 \frac{1}{\tfrac{1}{2}x - 1} \, dx$

## 15.3 Integration Needing Manipulation, Including Partial Fractions

**Exercise A**

1. (a) $\int \frac{x^2 + 2x + 2}{x+1} \, dx$  (b) $\int \frac{x(x+2)}{x+3} \, dx$  (c) $\int \frac{x^3 + 3x^2 + 5x + 3}{x^2 + x + 1} \, dx$

2. (a) $\int \sin x \sin 3x \, dx$  (b) $\int \cos x \cos 4x \, dx$

(c) $\int \sin 2x \cos x \, dx$  (d) $\int_0^{\pi/4} \cos 5x \cos 3x \, dx$

101

(e) $\displaystyle\int_{-\pi/6}^{\pi/6} \sin x \cos 3x\, dx$    (f) $\displaystyle\int_0^{\pi/6} \sin 2x \sin 4x\, dx$

3. (a) $\displaystyle\int_0^4 (\sin^2 x + \cos^2 x)\, dx$   (b) $\displaystyle\int_{\pi/6}^{\pi/4} (1 + \cot^2 x)\, dx$   (c) $\displaystyle\int_0^{\pi/3} (1 - 2\sin^2 x)\, dx$

(d) $\displaystyle\int \frac{\cos^2 x + \sin^2 x}{\cos^2 x}\, dx$   (e) $\displaystyle\int \sin^2 x\, dx$    (f) $\displaystyle\int \tan^2 x\, dx$

4. Integrate the following functions by first separating into partial fractions:

(a) $\dfrac{3x+4}{(x+1)(x+2)}$   (b) $\dfrac{5x-1}{(2x-1)(x+1)}$   (c) $\dfrac{9}{(2x+3)(x-3)}$

(d) $\dfrac{2x-9}{2x^2-3x}$   (e) $\dfrac{1}{2x^2+x-1}$   (f) $\dfrac{2}{2x^2+x-3}$

5. (a) $\displaystyle\int_1^3 \frac{x+1}{x^2+3x+2}\, dx$   (b) $\displaystyle\int_2^3 \frac{x+14}{x^2+3x-4}\, dx$

(c) $\displaystyle\int_2^3 \frac{1}{x^2+x-2}\, dx$   (d) $\displaystyle\int_2^3 \frac{x+1}{x^2+3x-4}\, dx$

(e) $\displaystyle\int_2^3 \frac{1}{(2x-3)(x+2)}\, dx$   (f) $\displaystyle\int_4^5 \frac{5-5x}{7x-3-2x^2}\, dx$

**Exercise B**

1. (a) $\displaystyle\int \frac{x^2+5x+7}{x+2}\, dx$   (b) $\displaystyle\int \frac{(x-1)(x-3)}{x-2}\, dx$   (c) $\displaystyle\int \frac{x^3+2x}{x^2-x+1}\, dx$

2. (a) $\displaystyle\int \sin 2x \cos 4x\, dx$   (b) $\displaystyle\int \cos 2x \cos 3x\, dx$

(c) $\displaystyle\int \sin x \sin 4x\, dx$   (d) $\displaystyle\int_0^{\pi/3} \sin x \sin 2x\, dx$

(e) $\displaystyle\int_0^{\pi/6} \cos 2x \cos 4x\, dx$   (f) $\displaystyle\int_0^{\pi/2} \sin x \cos 2x\, dx$

3. (a) $\displaystyle\int_0^{\pi/4} (\cos^2 x - \sin^2 x)\, dx$   (b) $\displaystyle\int_0^{\pi/3} (1 + \tan^2 x)\, dx$

(c) $\displaystyle\int_0^{\pi/3} (2\cos^2 x - 1)\, dx$   (d) $\displaystyle\int 2\sin x \cos x \cos 2x\, dx$

(e) $\displaystyle\int \cos^2 x\, dx$   (f) $\displaystyle\int \cot^2 x\, dx$

4. Integrate the following functions by first separating into partial fractions:

(a) $\dfrac{x+5}{(x-1)(x+2)}$   (b) $\dfrac{5x-4}{(3x-2)(x-1)}$   (c) $\dfrac{15}{(2x-5)(x+5)}$

(d) $\dfrac{7x-5}{2x^2-10x}$   (e) $\dfrac{1}{x^2+x-6}$   (f) $\dfrac{1}{6x^2-5x+1}$

5. (a) $\int_1^3 \dfrac{5}{2x^2+3x-2}\,dx$ (b) $\int_3^5 \dfrac{1}{x^2+3x-10}\,dx$

(c) $\int_6^8 \dfrac{2x+3}{17x-10-3x^2}\,dx$ (d) $\int_3^4 \dfrac{x-3}{x^2-3x+2}\,dx$

(e) $\int_1^3 \dfrac{1}{2x^2+x-1}\,dx$ (f) $\int_3^4 \dfrac{1}{11x-10-3x^2}\,dx$

## 15.4 Integration by Substitution

**Exercise A**

1. The following integrals may be solved by substitution, but the student should be able to do them by recognition using the principle

$$\int f'(u)\dfrac{du}{dx}\,dx = f(u) + C$$

(a) $\int (x^2+2)^5 2x\,dx$ (b) $\int \sin^6 x \cos x\,dx$ (c) $\int 2xe^{x^2}\,dx$

(d) $\int \dfrac{2x}{x^2-1}\,dx$ (e) $\int \tan^2 x \sec^2 x\,dx$ (f) $\int \dfrac{\cos x}{\sin x}\,dx$

(g) $\int x\sqrt{x^2+4}\,dx$ (h) $\int \cos x\, e^{\sin x}\,dx$ (i) $\int \dfrac{\sec^2 x}{\tan x}\,dx$

(j) $\int \sin 2x \cos^2 2x\,dx$

2. Evaluate the following integrals using the substitution suggested:

(a) $\int \dfrac{1}{(2x-1)^3}\,dx \quad (u=2x-1)$ (b) $\int \dfrac{dx}{\sqrt{1+x}} \quad (u^2=1+x)$

(c) $\int \dfrac{\sqrt{1+x^2}}{x}\,dx \quad (u^2=x^2+1)$ (d) $\int \dfrac{e^x}{e^x+e^{-x}}\,dx \quad (u=e^x)$

(e) $\int \sqrt{1-x^2}\,dx \quad (x=\sin\theta)$ (f) $\int \dfrac{\sqrt{x-1}}{x+1}\,dx \quad (x=u^2)$

(g) $\int \mathrm{cosec}\,\theta\,d\theta \quad (t=\tan\tfrac{1}{2}\theta)$ (h) $\int \dfrac{\cos\sqrt{x}}{\sqrt{x}}\,dx \quad (x=u^2)$

3. Solve the following by an appropriate substitution:

(a) $\int \dfrac{x^2}{(x+2)^4}\,dx$ (b) $\int \dfrac{1}{\sqrt{x+1}}\,dx$ (c) $\int \dfrac{1}{1+\sin\theta}\,d\theta$

(d) $\int \dfrac{e^{2x}}{e^x+1}\,dx$ (e) $\int \dfrac{1}{x\sqrt{x+1}}\,dx$ (f) $\int \dfrac{1}{12-13\sin\theta}\,d\theta$

(g) $\int \dfrac{x}{\sqrt{x^2+1}}\,dx$   (h) $\int \dfrac{x^3}{\sqrt{1-x^2}}\,dx$   (i) $\int \dfrac{dx}{(x-1)\sqrt{x}}$

**4.** Evaluate

(a) $\displaystyle\int_{5}^{10} \dfrac{x}{\sqrt{x-1}}\,dx$   (b) $\displaystyle\int_{1}^{4} \dfrac{\sqrt{x-1}}{\sqrt{x+1}}\,dx$   (c) $\displaystyle\int_{0}^{\sqrt{2}} \sqrt{4-x^2}\,dx$

(d) $\displaystyle\int_{0}^{\pi/2} \dfrac{1}{3+5\sin\theta}\,d\theta$   (e) $\displaystyle\int_{1}^{2} \dfrac{e^x}{e^{2x}-1}\,dx$   (f) $\displaystyle\int_{0}^{\pi/4} \dfrac{\cos 2x}{1+\sin 2x}\,dx$

### Exercise B

**1.** The following integrals may be solved by substitution but the student should be able to do them by inspection:

(a) $\int (x^3-1)^3 3x^2\,dx$   (b) $\int \cos^4 x \sin x\,dx$   (c) $\int 2x e^{-x^2}\,dx$

(d) $\int \dfrac{3x^2}{x^3-1}\,dx$   (e) $\int \cot x \operatorname{cosec}^2 x\,dx$   (f) $\int \dfrac{\sin x}{\cos x}\,dx$

**2.** (a) $\int \dfrac{e^x}{e^x+1}\,dx \quad (u=e^x)$   (b) $\int \dfrac{1}{1+\cos\theta}\,d\theta \quad (t=\tan\tfrac{1}{2}\theta)$

(c) $\int \dfrac{1}{(1+x^2)^{3/2}}\,dx \quad (x=\tan\theta)$   (d) $\int \dfrac{dx}{e^x-1} \quad (u=e^x)$

**3.** (a) $\int \dfrac{e^{2x}-1}{e^x-1}\,dx$   (b) $\int \dfrac{dx}{x\sqrt{x^2+1}}$   (c) $\int \dfrac{d\theta}{4\cos\theta-3\sin\theta}$

(d) $\int \dfrac{dx}{x+\sqrt{x}}$   (e) $\int \sin^5\theta\,d\theta$   (f) $\int \dfrac{1}{1-\cos 2\theta}\,d\theta$

**4.** (a) $\displaystyle\int_{1}^{2} \dfrac{dx}{\sqrt{5-2x}}$   (b) $\displaystyle\int_{0}^{\pi^2/4} \dfrac{\sin\sqrt{x}}{\sqrt{x}}\,dx$   (c) $\displaystyle\int_{0}^{1\tfrac{1}{2}} \dfrac{1}{\sqrt{9-x^2}}\,dx$

## 15.5 Integration by Parts

Evaluate the following integrals:

### Exercise A

**1.** $\int x\cos x\,dx$   **2.** $\int x e^x\,dx$   **3.** $\int x\sin 2x\,dx$   **4.** $\int x\log x\,dx$

**5.** $\displaystyle\int_{0}^{\pi/2} x\sin x\,dx$   **6.** $\displaystyle\int_{0}^{1} x e^{-x}\,dx$   **7.** $\displaystyle\int_{0}^{\pi/6} x\cos 3x\,dx$   **8.** $\displaystyle\int_{1}^{2} \dfrac{1}{x^2}\log x\,dx$

**9.** $\int x^2 \sin x\,dx$   **10.** $\int x^2 \log x\,dx$   **11.** $\int x^2 e^x\,dx$   **12.** $\int \dfrac{1}{x^2}\log\dfrac{1}{x}\,dx$

**13.** $\int \log x\,dx$   **14.** $\int \sin^{-1} x\,dx$   **15.** $\int x^3 e^x\,dx$

### Exercise B

1. $\int x \sin x \, dx$
2. $\int x e^{-x} \, dx$
3. $\int x \cos 2x \, dx$
4. $\int x e^{2x} \, dx$
5. $\int_0^{\pi/2} x \cos x \, dx$
6. $\int_1^2 x e^x \, dx$
7. $\int_0^{\pi/4} x \sec^2 x \, dx$
8. $\int_1^4 x \log x \, dx$
9. $\int x^2 \cos x \, dx$
10. $\int x^2 \sin \tfrac{1}{2} x \, dx$
11. $\int x^2 e^{-2x} \, dx$
12. $\int x^3 \log x \, dx$
13. $\int \log x^2 \, dx$
14. $\int \tan^{-1} x \, dx$
15. $\int x^3 e^{x^2} \, dx$.

## 15.6 Differential Equations

### Exercise A
Solve the following differential equations:

1. $x^2 \, dx = 2y \, dy$
2. $(x^2 + x) dx = y \, dy$
3. $\cos x \, dx = \sin y \, dy$
4. $\sec^2 x \, dx = 2y \, dy$
5. $\dfrac{dy}{dx} = \dfrac{2x+1}{2y+1}$
6. $\dfrac{2y+1}{x} \dfrac{dy}{dx} = \dfrac{1}{\sqrt{x^2-1}}$
7. $\dfrac{dy}{dx} = \dfrac{e^x}{y+1}$
8. $xy\sqrt{x^2+1} = \dfrac{1}{dx}$
9. $\dfrac{x}{1+x^2} = y \dfrac{dy}{dx}$
10. $(\sin x + \cos x)y = \dfrac{dy}{dx}$
11. $\dfrac{dy}{dx} = \cos^2 x \, e^{2x}$
12. $2y\sqrt{1-x^2} \dfrac{dy}{dx} = 1$
13. $(x+1)^2 \, dx = (y+1) dy$ if $y = 1$ when $x = 0$.
14. $\cos x \cos^2 y = \dfrac{dy}{dx}$ if $y = \tfrac{1}{4}\pi$ when $x = \tfrac{1}{6}\pi$
15. $\dfrac{dy}{dx} = \dfrac{y-1}{x-1}$ if $y = 3$ when $x = 2$.

### Exercise B
Solve the following differential equations:

1. $(x+1) dx = y^3 \, dy$
2. $(y+1)^2 \, dy = x \, dx$
3. $\sin x \, dx = \cos 2y \, dy$
4. $\operatorname{cosec}^2 y \, dy = 3x^2 \, dx$
5. $\dfrac{dy}{dx} = \dfrac{x^2}{(y+1)^2}$
6. $x\sqrt{x^2+9} = y^2 \dfrac{dy}{dx}$
7. $\dfrac{dy}{dx} = e^y(x^2+1)$
8. $xy(y^2+4) \dfrac{dy}{dx} = 1$
9. $\dfrac{2y}{y^2-1} \dfrac{dy}{dx} = 3x^2$
10. $\dfrac{dy}{dx} = e^{-y}(\cos x + \sec^2 x)$
11. $\dfrac{dy}{dx} = xye^x$
12. $\dfrac{dy}{dx} = \dfrac{y}{1+x^2}$
13. $\dfrac{dx}{(x+1)^2} = 3y^2 \, dy$ if $y = 1$ when $x = 0$.
14. $\dfrac{dy}{dx} + 2 \sin 2x \sec y = 0$ if $y = \tfrac{1}{2}\pi$ when $x = \tfrac{1}{2}\pi$.

105

15. $\dfrac{dy}{dx} = \dfrac{y}{x(x-1)}$ if $y = 2$ when $x = 2$.

16. $\dfrac{dy}{dx} = \cot y \cot x$ if $y = \tfrac{1}{3}\pi$ when $x = \tfrac{1}{2}\pi$.

## 15.7 Miscellaneous

1. $\displaystyle\int \dfrac{2}{2x-1}\,dx$ 
2. $\displaystyle\int \dfrac{2x}{2x-1}\,dx$ 
3. $\displaystyle\int \dfrac{2}{\sqrt{2x-1}}\,dx$

4. $\displaystyle\int_1^5 \dfrac{2x}{\sqrt{2x-1}}\,dx$ 
5. $\displaystyle\int_0^4 x\sqrt{4-x}\,dx$ 
6. $\displaystyle\int x\sqrt{4-x^2}\,dx$

7. $\displaystyle\int_0^{\sqrt{2}} \sqrt{4-x^2}\,dx$ 
8. $\displaystyle\int_0^1 x^2\sqrt{4-x^2}\,dx$ 
9. $\displaystyle\int x\sqrt[3]{x^2+1}\,dx$

10. $\displaystyle\int \dfrac{x^3+x^2+x+2}{x^2+1}\,dx$ 
11. $\displaystyle\int \sin^2 2x \cos^2 2x\,dx$ 
12. $\displaystyle\int \dfrac{1}{e^{-x}+1}\,dx$

13. $\displaystyle\int_4^{11} \dfrac{x+12}{x^2+3x-10}\,dx$ 
14. $\displaystyle\int_0^{\pi/2} \sin 3x \sin 4x\,dx$ 
15. $\displaystyle\int \dfrac{\cos x}{\operatorname{cosec}^3 x}\,dx$

16. $\displaystyle\int \dfrac{x-1}{x^2+1}\,dx$ 
17. $\displaystyle\int \dfrac{x(x+1)}{x-1}\,dx$ 
18. $\displaystyle\int \cos 3x \cos 4x\,dx$

19. $\displaystyle\int_0^{\pi/2} \sqrt{1+\cos x}\,dx$ 
20. $\displaystyle\int x \cos 4x\,dx$ 
21. $\displaystyle\int \sin 3x \cos 4x\,dx$

22. $\displaystyle\int \dfrac{2e^{2x}+e^x}{e^{2x}+e^x}\,dx$ 
23. $\displaystyle\int \dfrac{x}{(x^2+1)\sqrt{x^2+1}}\,dx$ 
24. $\displaystyle\int x \sin x \cos x\,dx$

25. $\displaystyle\int \dfrac{1}{(x^2+2x+1)\sqrt{x+1}}\,dx$ 
26. $\displaystyle\int \dfrac{dx}{\sqrt{3-5x}}$ 
27. $\displaystyle\int_0^3 \dfrac{\sqrt{x}}{x+1}\,dx$

28. $\displaystyle\int \dfrac{dx}{x^2-x}$ 
29. $\displaystyle\int \sin x \cos x \cos 3x\,dx$ 
30. $\displaystyle\int \dfrac{1+e^x}{e^{-x}+e^x}\,dx$

31. Use the substitution $x = \tan\theta$ to show that
$$\int_0^1 \dfrac{x^2}{(1+x^2)^{3/2}}\,dx = \int_0^{\pi/4} (\sec\theta - \cos\theta)\,d\theta$$
and hence evaluate the integral.

32. If $y = \tan\theta + \sec\theta$ and $x = \cot\theta + \operatorname{cosec}\theta$ show that
$$\dfrac{x^2+1}{x} = 2\operatorname{cosec}\theta \quad \text{and} \quad \dfrac{y^2+1}{y} = 2\sec\theta.$$
Hence show that
$$\dfrac{dy}{dx} = -\dfrac{1+y^2}{1+x^2}.$$

Solve this differential equation and show that your result is equivalent to $\frac{x+y}{1-xy}$ = constant. Use any convenient value of $\theta$ to evaluate the constant.

33. A particle falls from rest through air which is assumed to have a resistance which varies directly as the velocity. Show that the equation of motion is given by $\frac{dv}{dt} = g - kv$. Solve this differential equation and show that the velocity never exceeds $\frac{g}{k}$. Find also the distance fallen in $t$ seconds.

34. If in question 33 the resistance is taken to vary as the square of the velocity then the equation is $\frac{dv}{dt} = g - kv^2$. Show that in this case the velocity is given by $v = a\frac{e^{2akt} - 1}{e^{2akt} + 1}$, where $g = a^2k$. Find the 'terminal velocity', i.e. the maximum velocity when the time is increased without limit.

35. A tangent is drawn to a curve at a point $P$ on the curve cutting the axes at $A$ and $B$. If $P$ is found to be the mid-point of $AB$ show that $\frac{dy}{dx} = -\frac{y}{x}$. Find the equation of the curve if it passes through the point (4, 9).

# 16

# Applications of Integration

## 16.1 Notes and Formulae

Area under $y = f(x)$ between $x = a$ and $x = b$ is $\int_a^b y\,dx$.

Area between $x = g(y)$, the $y$-axis, $y = c$ and $y = d$ is $\int_c^d x\,dy$.

Volume of solid of revolution about $x$-axis is $\int_a^b \pi y^2 \,dx$.

Volume of solid of revolution about $y$-axis is $\int_c^d \pi x^2 \,dy$.

Centroid of plane area $\bar{x} = \int_a^b xy\,dx \Big/ \int_a^b y\,dx.$

$$\bar{y} = \int_a^b \tfrac{1}{2}y^2\,dx \Big/ \int_a^b y\,dx.$$

In all cases of area, volumes, and centre of gravity the student is advised to sketch the curve. For curves which are not single valued functions some modification of the formulae may be necessary.

Centre of Gravity of solid of revolution about x-axis:

$$\bar{x} = \int_a^b \pi xy^2\,dx \Big/ \int_a^b \pi y^2\,dx.$$

Mean value of $y = f(x)$ from $x = a$ to $x = b$ is $\int_a^b y\,dx/(b-a)$.

Trapezoidal rule: $\int y\,dx \approx \dfrac{h}{2}\{y_1 + y_n + 2(y_2 + y_3 + \ldots + y_{n-1})\}$

Simpson's rule $\int y\,dx \approx \dfrac{h}{3}\{y_1 + y_n + 4(\text{sum of even ordinates} + 2(\text{sum of other odd ordinates})\}$

It is easier to use the simple formula $\dfrac{h}{3}(y_1 + y_3 + 4y_2)$ for each pair of strips. In any case there must be an odd number of ordinates for Simpson's rule.

In such cases, it is advisable to make an estimate of the answer so as to avoid simple errors. For instance it is usually possible to see that a given area must be less than a circumscribed rectangle, but greater than an inscribed triangle.

## 16.2 Areas

1. Find the areas between the following curves and lines:
   (a) $y = 4x + 2$, $y = 0$, $x = 1$ and $x = 2$.
   (b) $3y - x = 6$, $y = 0$, $x = -3$ and $x = 3$.
   (c) $y = x^2 - 4$ and $y = 0$.
   (d) $y = \dfrac{1}{x^2}$, $y = 0$, $x = 2$ and $x = 4$.
   (e) $y = \cos 3x + \sin x$, $y = 0$, $x = 0$ and $x = \dfrac{\pi}{2}$.
   (f) $x = 3 - y^2$, $x = 0$, $y = 0$ and $y = 1$.
   (g) $x - y^3 = y^2$, $x = 0$ and $y = 1$.
2. Find the area between the following curves and lines:
   (a) $y = 4x^2$ and $y = 36$.
   (b) $x = y^2 - 1$ and $x = 3$.
   (c) $y = 2x^2$ and $y = x + 1$.
   (d) $y = (x+1)(x+2)$ and $y - x = 2$.
   (e) $y = 4x^2$ and $2y = 2x + 3$.
   (f) $y = x^3 + x$ and $y = 2x^3$.
   (g) $y^2 = 4x$ and $y^2 = -4(x-2)$.

3. Find the ratio in which the line $y = 2x$ divides the area enclosed by the curve $y = 8x - x^2$ and the x-axis.
4. Find the area enclosed by the curve $y = \sin x$ and the line $2y - 1 = 0$. $(0 \leqslant x \leqslant \pi)$.
5. Find the area bounded by the curve $y = e^{2x-1}$, the x-axis and the lines $x = 1$ and $x = 4$.
6. Differentiate $x \log_e x$ and use your result to evaluate $\int \log_e x \, dx$. Use this result to find the area enclosed by the curve $y = \log_e x$, the x-axis and the line $x = e$. Similarly find the area enclosed by $y = \log_e x$, the x-axis and the line $x = 10$.
7. Find the area enclosed between the curve $y^2 = x^3$ and the line $x = 4$.
8. Show that one of the points of intersection of the curve $xy = 12$ and the curve $y = 7x - 4 - x^2$ lies on the line $x = 2$. Find both points of intersection and calculate the area enclosed between the two curves in the first quadrant.
9. Sketch the curves $y = 6x - x^2$ and $y - 1 = (x - 3)^2$. Find their points of intersection and calculate the area enclosed between them.
10. Find the area between the curve $y = \dfrac{1}{x-1} + 1$, the x-axis and the ordinates $x = 2$ and $x = 6$.
11. Sketch the curve $y = x^3 - 6x^2 + 11x - 6$ and show that this curve and the x-axis enclose two equal areas.
12. Find the area between the curve $pv^{1.41} = k$, the v-axis and the lines $v = v_1$ and $v = v_2$. ($k$ is a constant.)

## 16.3 Volumes

(Leave $\pi$ in answers.)
1. Find the volumes of the solids generated by rotating the following curves about the axes shown in brackets and between the limits given:
   (a) $y = 2x + 4$ from $x = 0$ to $x = 5$ (x-axis).
   (b) $y^2 = x - 1$ from $y = -1$ to $y = 1$ (y-axis).
   (c) $2y - \sqrt{x} = 2$ from $x = 1$ to $x = 9$ (x-axis).
   (d) $x^2 - y^3 = y$ from $y = 1$ to $y = 2$ (y-axis).
   (e) $y^2 = (x + 1)(4 - x)$ from $x = -1$ to $x = 4$ (x-axis).
   (f) $y^2 = \dfrac{1}{x}$ from $y = 1$ to $y = 2$ (y-axis).
   (g) $y = 2 \sin x$, $x = 0$, $x = \frac{1}{4}\pi$ (x-axis).
2. Find the volumes generated when the areas between the following curves and lines are rotated about the lines shown in brackets:
   (a) $y = 4 - x^2$, $y = 0$ (x-axis).
   (b) $y^2 = 4x$, $x = 4$ (x-axis).
   (c) $y^2 = 4x$, $y = 0$, $y = 4$ (y-axis).
   (d) $y^2 = 4x$, $y = 0$, $x = 4$ (y-axis).
   (e) $y^2 = \sin x$, $x = 0$, $x = \pi$, $y = \dfrac{1}{\sqrt{2}}$ (x-axis).
   (f) $y = x^2 - 4x + 2$, $y = 2$ ($y = 2$).
   (g) $y = \sqrt{x}$, $y = 0$, $x = 2$ ($x = 2$).

109

3. The area enclosed in the first quadrant between the curve $4x^2 + y^2 = 4$ and the axes is rotated about the x-axis through $360°$. Find the volume of the resulting solid.
4. Find the volume of the solid formed when the portion of the curve $y = \sec x$ between $x = 0$ and $x = \frac{1}{4}\pi$ radians is rotated about the x-axis.
5. The area enclosed by the curve $y = x^2(4-x)$ and the x-axis is rotated about the x-axis. Find the volume of the solid so formed.
6. A wineglass is made in the shape formed by rotating the curve $y = x^2$ about the y-axis and is such that the diameter of the rim is 6 cm and the volume is $40\frac{1}{2}\pi$ cm$^3$. Find the height of the bowl.
7. The smaller area enclosed between the two curves $x^2 + y^2 = 25$ and $4y^2 = 9x$ is rotated about the x-axis. Find the volume of the figure so formed.
8. If the area between the curve $y = 1 + \sin x$, the x-axis, the y-axis and the ordinate $x = \pi$ is rotated about the x-axis show that the volume so formed is $\frac{1}{2}\pi(3\pi + 8)$.
9. A spherical cap is cut off a sphere of radius $r$ by a plane distant $c$ from the circumference. Prove that the volume of this cap is $\frac{1}{3}\pi c^2(3r - c)$.

## 16.4 Centroids and Centres of Gravity

1. Find the co-ordinates of the centroids of the areas bounded by the following curves and lines:
   (a) $y = 3x + 4$, $x = 0$, $x = 2$, $y = 0$.
   (b) $y^2 = 4x$, $x = 4$.
   (c) $x^2 = 4 - y$, $x = 0$.
   (d) $y = 6x - x^2$, $y = 0$.
   (e) $y = 2x^3 + 1$, $y = 1$, $x = 3$.
2. Find the centre of gravity of the solids of revolution which are generated by rotating the areas between the following curves and lines about the axes given in brackets:
   (a) $y = x$, $y = 0$, $x = 4$ (x-axis).
   (b) $y = 4x$, $x = 4$, $x = 2$, $y = 0$ (x-axis).
   (c) $y^2 = x$, $x = 0$, $y = 2$ (y-axis).
   (d) $y = (x+3)(x-2)$, $y = 0$ (x-axis).
   (e) $xy = 4$, $y = 0$, $x = 2$, $x = 4$ (x-axis).
3. Find the centroid of the area formed between the curve $x^2 + y^2 = 25$ and the two axes in the first quadrant.
4. Find the centre of gravity of a cone. $\left(\text{Consider the line } y = \dfrac{r}{h}x.\right)$
5. Sketch the curve $y^2 = x(x-4)^2$. Find
   (a) the area of the loop.
   (b) the centroid of the loop.
   (c) the volume of the solid formed by rotating the loop about the axis of x.
   (d) the centre of gravity of the solid defined in (c).
6. Sketch the curve $x = 5y - y^2 - 6$ and find the centroid of the area bounded by this curve and the y-axis.
7. Find the centre of gravity of a solid hemisphere.
8. Sketch the curve $y^2 = x^2(9 - x^2)$. Find

(a) the area enclosed by one loop.
(b) the centroid of this area.
(c) the volume of the solid formed by rotating one loop about the x-axis through $\pi$ radians.

## 16.5 Mean Values

**Exercise A**
Find the mean value of the following functions over the intervals stated:
1. $4x + 5$ over the interval 0 to 4
2. $x^3 - 3x^2 + 5$           $-2$ to $+2$
3. $\dfrac{1}{1+x^2}$           1 to $\sqrt{3}$
4. $\sin^2 x$           0 to $\frac{1}{2}\pi$
5. $\tan x$           0 to $\frac{1}{4}\pi$

**Exercise B**
Find the mean values of the following functions over the intervals stated:
1. $(x+4)(2x-3)$ over the interval 2 to 3
2. $48 - 3x^2$           0 to 4
3. $\sin x$           0 to $\pi$
4. $xe^{x^2}$           1 to 2
5. $\dfrac{x}{x^2+1}$           1 to 3

## 16.6 Approximate Integration

**Exercise A**
1. Use the trapezoidal rule to estimate the area under the following curves between the limits given:
   (a) $y = \sqrt{x^2+1}$,    $x = 0$ to $x = 1$    (11 ordinates)
   (b) $y = e^{x^2}$,    $x = 1$ to $x = 2$    (6 ordinates)
   (c) $y = \log_e(\sin x)$,    $x = \frac{1}{12}\pi$ to $x = \frac{1}{2}\pi$    (6 ordinates)
2. Use Simpson's rule to estimate the area under the following curves between the limits given:
   (a) $y = \dfrac{1}{x}$,    $x = 4$ to $x = 5$    (9 ordinates)
   (b) $y = \dfrac{1}{1+x^2}$,    $x = 0$ to $x = 1$    (5 ordinates)
   (c) $y = \tan^2 x$,    $x = 0$ to $x = \frac{1}{3}\pi$    (5 ordinates)
3. Evaluate the following integrals by approximate integration using the method given:
   (a) $\displaystyle\int_1^2 \dfrac{1}{x}\, dx$    (Trapezoidal rule: 11 ordinates.)
   (b) $\displaystyle\int_0^{\pi/2} \sqrt{\sin x}\, dx$    (Simpson's rule: 5 ordinates.)

(c) $\int_0^3 x(1+\sqrt{x})^{1/2} dx$ (Simpson's rule: 7 ordinates.)

**Exercise B**

1. Use the Trapezoidal rule to estimate the area under the following curves between the limits given:
   (a) $y = (x+2)(x+3)$    $x = 0$ to $x = 2$    (11 ordinates)
   (b) $y = \sin 3x$    $x = 0$ to $x = \frac{1}{4}\pi$    (5 ordinates)
   (c) $y = e^{\cos x}$    $x = 0$ to $x = \frac{1}{2}\pi$    (5 ordinates)

2. Use Simpson's rule to estimate the area under the following curves between the limits given:
   (a) $y = \sin^2 x$    $x = 0$ to $x = \frac{1}{2}\pi$    (7 ordinates)
   (b) $y = e^{-2x^2}$    $x = 0$ to $x = 2$    (5 ordinates)
   (c) $y = \log_{10}(10 \sin x)$    $x = \frac{1}{6}\pi$ to $x = \frac{1}{2}\pi$    (5 ordinates)

## 16.7 Miscellaneous

1. Sketch the curve $y^2 = x^3$. Find
   (a) the area enclosed by the curve and the line $x = 4$,
   (b) the centroid of this area,
   (c) the volume of the solid formed by rotating the area about the axis of $x$.
   (d) the centre of gravity of this solid,
   (e) the volume of the solid formed by rotating about the $y$-axis the area enclosed by the curve, the line $y = 8$ and the $y$-axis.
   (f) the centre of gravity of the solid defined in (e).

2. Sketch the curve $y = e^{-x^2}$ and use Simpson's rule to show that the area under this curve between $x = 0$ and $x = 2$ is approximately 0.8818. Find the co-ordinates of the centroid of the area bounded by the curve, the $x$-axis and the ordinates $x = \pm 2$.

3. Sketch the graph of $y = 8x - x^2$ and find the area enclosed by the curve and the $x$-axis:
   (a) by using the trapezoidal rule with 5 ordinates (over half the area),
   (b) by using Simpson's rule with the same strips,
   (c) by normal integration.

4. The parabola $y^2 = 8x$ is rotated about the $x$-axis through two right angles. Find the centre of gravity of the solid cut off by the plane $x = 8$.

5. Employ Simpson's rule to evaluate $\int_0^6 (x^3 - 14x^2 + 48x) dx$ by using 7 ordinates. Compare your result with that obtained by normal integration and comment on the result. (The same identity of results will occur with any cubic.)

6. Sketch the ellipse $9x^2 + 16y^2 = 144$ and the hyperbola $9x^2 - 4y^2 = 9$. The area in the first quadrant bounded by the two curves and the $x$-axis is rotated through four right angles about the $x$-axis. Prove that the volume of the solid is $10\frac{1}{2}\pi$.

7. Sketch the curves $y^2 = 4ax$ and $y^2 = 4bx$ where $b > a$. Shade the area in the first quadrant between the two curves and the lines $x = a$ and $x = b$. This area is rotated about the x-axis through $2\pi$ radians. Prove that the volume of the solid formed is $2\pi(b-a)^2(a+b)$.
8. Sketch the curve $xy = 3$ and the straight line $x + y = 4$ on the same axes. Find the area enclosed by the curve and the line and also the volume of the figure formed by rotating the given area about the x-axis.

# Answers

## 1.2

### Exercise A

1 3125, 2401, 1024, $-\dfrac{1}{128}$, $-\dfrac{1}{243}$, 1   2 $\pm 9$, 4, $\pm 3$, $\pm 15$, 9, 1
3 $\tfrac{1}{10}$, $\tfrac{1}{25}$, $\tfrac{1}{81}$, 1, 32, $\pm 5$   4 6, 16, 1, $\pm 27$, $\tfrac{1}{16}$, $\pm \tfrac{7}{5}$
5 2.828, 3.464, 4.472, 0.707, 0.4472   6 3, 4, 2, 4,   7 $1\tfrac{1}{2}$, $-3$, $-2\tfrac{1}{2}$, $-3$, 0, $1\tfrac{1}{2}$, $1\tfrac{1}{4}$, $-2\tfrac{1}{2}$   8(a) 16   (b) 8   (c) 3 or 27   (d) 8 or $4\tfrac{2}{3}$
(e) 9   9(a) (4, 16)   (b) (16, 8)   (c) (32, 125)   (d) (2, 8)
(e) $(1\tfrac{1}{2}, 4\tfrac{1}{2})$ or $(-1, 2)$   (f) $(27, \tfrac{1}{3})$

### Exercise B

1 5.656, 4.898, 5.196, 0.5773, 0.4082   2 3, 3, 2, 5   3 $1\tfrac{1}{2}$, $-4$, $-2$, $-3$   4 0, $-2\tfrac{1}{2}$, $1\tfrac{1}{4}$, $-3$   5(a) 9   (b) 25   (c) 8 or $\tfrac{1}{8}$   (d) 16 or $\tfrac{1}{16}$
(e) $x = 2$ or 64   6(a) $(27, \tfrac{1}{3})$   (b) (27, 16)   (c) $(125, \tfrac{1}{4})$   (d) $(\tfrac{1}{2}, 1)$
(e) (4, 4)   (f) (8, 16)

## 1.3

### Exercise A

1 $\dfrac{2}{x-1} + \dfrac{3}{x+2}$   2 $\dfrac{3}{x-3} - \dfrac{2}{x-5}$   3 $\dfrac{4}{x-2} + \dfrac{5}{x+5}$
4 $\dfrac{2}{2x-3} + \dfrac{5}{3x+2}$   5 $\dfrac{7}{2x+1} - \dfrac{3}{2x-1}$   6 $\dfrac{2}{5(2x-3)} - \dfrac{1}{5(x+1)}$
7 $\dfrac{1}{9(2x-1)} + \dfrac{4}{9(x+4)}$   8 $\dfrac{1}{5(x-3)} - \dfrac{3}{5(3x+1)}$

113

**9** $\dfrac{3}{7(x-1)} + \dfrac{15}{7(2x+5)}$   **10** $\dfrac{1}{x+3} + \dfrac{2}{x-2} - \dfrac{2}{x+1}$   **11** $\dfrac{x+5}{2x^2+5} - \dfrac{1}{2x-1}$

**12** $\dfrac{2}{2x-3} - \dfrac{x-5}{x^2+5}$   **13** $\dfrac{3}{(x+2)^2} - \dfrac{2}{x+2} + \dfrac{2}{x-1}$

**14** $\dfrac{1}{(x-3)^3} - \dfrac{2}{(x-3)^2} + \dfrac{3}{x-3}$

### Exercise B

**1** $\dfrac{1}{x-2} - \dfrac{3}{x+3}$   **2** $\dfrac{4}{x+1} - \dfrac{3}{x+2}$   **3** $\dfrac{3}{x-9} - \dfrac{4}{2x-1}$

**4** $\dfrac{11}{3(3x-5)} - \dfrac{10}{3(3x+5)}$   **5** $\dfrac{3}{2x-1} - \dfrac{2}{3x+1}$   **6** $\dfrac{1}{9(x-2)} - \dfrac{2}{9(2x+5)}$

**7** $\dfrac{1}{5(x-1)} + \dfrac{2}{5(3x+2)}$   **8** $\dfrac{1}{5(2x-1)} + \dfrac{1}{5(3x+1)}$

**9** $\dfrac{2}{x-3} + \dfrac{3}{x+2} - \dfrac{1}{x+4}$   **10** $\dfrac{3x-2}{3x^2-1} - \dfrac{1}{x+5}$   **11** $\dfrac{2}{x-1} - \dfrac{2x-1}{x^2+x+1}$

**12** $\dfrac{3x-2}{x^2+2x+3} - \dfrac{3}{x-5}$   **13** $\dfrac{1}{(x-3)^2} - \dfrac{3}{x-3} + \dfrac{3}{x+4}$

**14** $\dfrac{2}{(2x-1)^3} - \dfrac{1}{(2x-1)^2} + \dfrac{3}{2x-1}$

## 1.4

### Exercise A

**1** $x^2 + y^2 = a^2$   **2** $xy = 1$   **3** $\dfrac{x^2}{a^2} - \dfrac{y^2}{b^2} = 1$   **4** $y = \dfrac{x}{a} + \dfrac{a}{x}$

**5** $y = \dfrac{2x}{a}\sqrt{a^2 - x^2}$   **6** $y = 3(x-1)^2 - 2(x-1)^3$   **7** $x^{2/3} + y^{2/3} = a^{2/3}$

**8** $y = 2x - 1$   **9** $ay = 2x^2 - a^2$   **10** $4(y+2)^2 + 9(x-3)^2 = 36$

**11** $x + 2y = 3$   **12** $x^2y^2 + a^2b^2 = a^2y^2$   **13** $y = x^2 + 2x$

**14** $x^2 - y^2 = 4$

### Exercise B

**1** $x^2 - y^2 = a^2$   **2** $b^2x^2 + a^2y^2 = a^2b^2$   **3** $xy = c^2$   **4** $b^2x = ay^2$

**5** $2y^2 + x = 1$   **6** $x = 2y + 5$   **7** $x^2 + y^2 - 2x - 2y + 1 = 0$

**8** $y = x^2 + 4x$   **9** $x^2 = y^2 + 4$   **10** $(x+1)(y+3) = 6$   **11** $y^2 = e^2x$

**12** $p^2y^2 = x^2y^2 + q^2x^2$   **13** $x = (y+1)^3 - 1$   **14** $x^2 = y + 2$

## 1.6

**2**(a) $1\tfrac{3}{4} \log_2 x$   (b) $\log_2 x + \tfrac{1}{2}$   **4** $1 + \dfrac{2}{2x+1} - \dfrac{1}{x-1}$

(a) $\dfrac{16}{(2x+1)^3} - \dfrac{2}{(x-1)^3}$   (b) $x + \log_e \dfrac{2x+1}{x-1} + c$   5 $\dfrac{y}{x} = 4$ or $-1$

8  $x + 5y + 9 = 0$   9  $x^4 - x^2 y^2 + 2y^4 = 0$

## 2.2

### Exercise A
1(a) $-4, \tfrac{1}{2}$   (b) 6.32 or $-0.317$   (c) unreal   (d) 1.27 or $-0.472$
(e) unreal   (f) 1 or $-\tfrac{1}{3}$   2(a) real and rational   (b) real and equal
(c) real and irrational   (d) unreal   (e) real and equal   (f) real and irrational   (g) real and rational   (h) unreal   (i) real and equal
3(a) $\pm 10$   (b) $\pm 2$ or $\pm 14$   (c) $\pm 7b$ or $\pm 5b$   (d) 6   4(a) $b = 0$
(b) $c = a$   5(a) $a \leqslant -10$ or $a \geqslant 10$   (b) $-8 < a < 8$   (c) all real values of $a$   6(a) $-1$   (b) 2   (c) 2 or $-\dfrac{13}{7}$   (d) $\pm 2$

7(a) 1   (b) $-3$   8(a) $y \geqslant -\dfrac{17}{8}$   (b) all real values   (c) $y < -1$ or $y > \tfrac{3}{2}$   (d) $y \leqslant -40 - \sqrt{1560}$ or $y \geqslant -40 + \sqrt{1560}$
9(a) $x < -\tfrac{4}{3}$ or $x > 2$   (b) $x < -\tfrac{1}{2}$ or $x > 1$   (c) $-1 < x < 2\tfrac{1}{2}$
(d) $x < \dfrac{2}{a-2}$ or $x > 1$   (e) $x < -3$ or $\tfrac{1}{2} < x < 2$   (f) $x < -2\tfrac{1}{2}$, $-2 < x < -\tfrac{2}{3}$, $x > 2$

### Exercise B
1(a) real and rational   (b) real and irrational   (c) unreal   (d) real and rational   (e) unreal   (f) unreal   (g) unreal   (h) real and rational
(i) real and rational   2(a) $\pm 20$ or $\pm 4$   (b) 1   (c) 25   (d) 0 or 4
3(a) $q^2 = 4pr$   (b) $q = 0$   4(a) $-10 < p < 10$   (b) $p \leqslant 1$   (c) $p \leqslant \tfrac{1}{3}$
5(a) 5 or $-1$   (b) 3   (c) 7   6(a) $-\tfrac{7}{2}$   (b) $-1 \pm \sqrt{6}$   7(a) $y \leqslant 1$
(b) $y \leqslant -4$ or $y \geqslant 2$   (c) all real values   (d) $y \leqslant -40 - \sqrt{1640}$ or $y \geqslant \sqrt{1640} - 40$   8(a) $x < -\tfrac{2}{3}$ or $x > 1\tfrac{1}{2}$   (b) $-1\tfrac{1}{2} < x < 1$
(c) $x < -1$ or $x > 4$   (d) $-1 < x < -\tfrac{1}{3}$ or $x > 5$   (e) $x < 1\tfrac{1}{2}$ or $x > 4$

## 2.3

### Exercise A
1(a) 7   (b) 5   (c) 18   (d) $8\sqrt{5}$ if $\alpha > \beta$ or $-8\sqrt{5}$ if $\alpha < \beta$   2(a) $-1$
(b) $-\tfrac{1}{3}$   (c) $-2\tfrac{1}{4}$   (d) $-2\tfrac{1}{4}$   3(a) 1   (b) $\dfrac{a(2a^2 - 3)}{(a^2 - 1)^2}$
(c) $a(a^2 - 2)\sqrt{4 - 3a^2}$   4(a) $x^2 + 4x + 12 = 0$
(b) $x^2 + 2x + 9 = 0$   (c) $3x^2 + 2x + 1 = 0$   (d) $x^2 - 10x + 27 = 0$
5(a) $6x^2 + 11x + 6 = 0$   (b) $9x^2 + 11x + 4 = 0$   (c) $2x^2 - 5x + 6 = 0$
(d) $2x^2 - 3x + 4 = 0$

### Exercise B

**1**(a) 17    (b) 21    (c) $-63$    (d) $-\dfrac{17}{4}$    **2**(a) $\dfrac{22\sqrt{76}}{27}$    (b) $-\dfrac{29}{54}$
(c) $-\dfrac{80}{9}$    (d) $-\dfrac{476}{81}$    **3**(a) $p(p^3 - 3)$    (b) $-p^2\sqrt{p(p^3-4)}$
(c) $p^3(2-p^3)\sqrt{p(p^3-4)}$    **4**(a) $x^2 + 3x - 1 = 0$    (b) $x^2 - 11x + 1 = 0$
(c) $x^2 + x - 3 = 0$    **5**(a) $64x^2 - 24x + 27 = 0$    (b) $12x^2 + 16x + 9 = 0$
(c) $x^2 + x + 3 = 0$    (d) $48x^2 - 28x + 27 = 0$

## 2.4

### Exercise A

**1**(a) $\left(-\dfrac{5}{11}, -\dfrac{26}{11}\right)$ or $(1, 2)$    (b) $(3, 1)$ or $(-8, 6\tfrac{1}{2})$    (c) $(3, 2)$ or $(-23, -37)$    (d) $(-2, -1)$ or $\left(\dfrac{8}{5}, \dfrac{1}{5}\right)$    **2** $(3, 9)$ and $(1, 5)$    **3** $(3, 4)$
**5** $\left(\dfrac{69}{25}, \dfrac{7}{5}\right)$    **6**(a) $-1 < a < \dfrac{17}{8}$    (b) $a = 1$    (c) $a = \dfrac{17}{8}$ or $-1 < a < 1$
(d) $-1$    (e) $a > \dfrac{17}{8}$ or $a < -1$    **7** $(2b - 7)^2 = 16a(c - 2)$

### Exercise B

**1**(a) $(4, 2)$    (b) $(2, 1)$ or $(7, 11)$    (c) $(4.764, 5.292)$ or $(2.036, -2.892)$
(d) $(2.913, -1.870)$ or $(0.905, 1.143)$    **2** $(8, 8)$ and $(2, -4)$    **3** $(2, 7)$
**5** $(2\tfrac{1}{4}, 2\tfrac{1}{2})$    **6** $\left(\dfrac{32}{25}, -\dfrac{18}{25}\right)$    **7** $a = \pm\sqrt{5}$

## 2.5

**1** 12.0, 0.00417    **3** $-2$; $-1, 3$ and $0, 4$    **5** 6 or $\dfrac{14}{9}$    **8** no values
**9** $\alpha = \beta = -1$; $p = 1, q = 0$    **10** $a = -2, b = -9$

## 3.2

**1**(a) 720    (b) 5040    (c) 120    (d) 3628800    **2**(a) 30    (b) 6720
(c) 40320    (d) 970200    (e) 9900    **3**(a) A, B, C, D    (b) AB, AC, AD, BA, BC, BD, CA, CB, CD, DA, DB, DC    (c) ABC, ABD, ADB, ACB ACD, ADC, BAC, BCA, BAD, BDA, BCD, BDC, CAB, CBA, CAD, CDA, CBD, CDB, DAB, DBA, DAC, DCA, DBC, DCB    **4** $6! = 720$    **5** 720    **6** 60    **7** 24360
**8**(a) 336    (b) 1680    **9** 120    **10** 32760    **11**(a) 2520    (b) 120
(c) 5040    (d) 840    (e) 60    (f) 1630    (g) 3465    **12** 360
**13**(a) ABC, ACB, BCA, BAC, CAB, CBA, CAC, CBC, ACC, BCC, CCA, CCB, CCC    (b) ABCC, ACBC, ACCB, ACCC, BACC, BCAC, BCCA,

BCCC, CABC, CACB, CCAB, CBAC, CBCA, CCBA, CCCA, CCAC, CACC, CCCB, CCBC, CBCC   (c) ABCCC, BACCC, CABCC, CBCAC, CACCB, ACBCC, BCACC, CACBC, CCBAC, CCCBA, ACCBC, BCCAC, CCABC, CCCAB, CCBCA, ACCCB, BCCCA, CBACC, CCACB, CBCCA   **14**(a) 72   (b) 12   **15** 2520   **16** 180, 60   **17**(a) 2!4!   (b) 10!12!4!   (c) 8!12!   (d) 8!

## 3.3

**1**(a) $\dfrac{4!}{2!2!}$   (b) $\dfrac{9!}{7!2!}$   (c) $\dfrac{6!}{2!4!}$   (d) $\dfrac{6!}{4!2!}$   (e) $\dfrac{11!}{9!2!}$   **2**(a) 10   (b) 15
(c) 126   (d) 126   (e) 120   (f) 1   **3**(b) 6   (d) 10   **4** 1140
**5** 480700   **6** 35   **7** 220   **8** 30045015   **9** 792   **10**(a) 1716
(b) 924   **11**(a) 63   (b) 42   **12** 31   **13** 127   **14** 700   **15**(a) 4096
(b) 3937   **16**(a) 9   (b) 10

## 3.4

**1** 12   **2**(a) 432   (b) 810   **3** 608400   **4**(a) 10   (b) 6   **5** 729
**6**(a) 144   (b) 30   **7**(a) 81   (b) 118   (c) 1459   (d) 18, 27, 325
**8**(a) 15   (b) 2   (c) 7   **9** 20   **10** 60   **11**(a) 2520   (b) 504
**12**(a) 56   (b) 336   **13**(a) 10, 42   (b) 63, 357   (c) 6, 19
**14**(a) 720   (b) 60   **15** 5040   **16** 455   **17** 204   **18** 628

## 3.5

### Exercise A

**1**(a) $\dfrac{3}{13}$   (b) $\dfrac{7}{8}$   (c) $\dfrac{1}{2}$   (d) $\dfrac{1}{6}$   (e) $\dfrac{73}{216}$   (f) $\dfrac{5}{21}$   **2**(a) $\dfrac{12}{65}$   (b) $\dfrac{27}{91}$
(c) $\dfrac{53}{65}$   (d) $\dfrac{236}{455}$   (e) $\dfrac{27}{91}$   (f) 0   (g) $\dfrac{7}{13}$   **3**(a) $\dfrac{5}{22}$   (b) $\dfrac{1}{11}$   **4** $\dfrac{11}{26}$
**5** $\tfrac{1}{4}\pi$   **6**(a) $\dfrac{3}{5}$   (b) $\dfrac{5}{9}$   (c) $\dfrac{2}{5}$   **7**(a) $\dfrac{86}{200}$   (b) $\dfrac{1}{64}$   (c) $\dfrac{729}{10000}$   **8** $\dfrac{315}{391}$
**9**(a) $\dfrac{2}{5}$   (b) $\dfrac{1}{5}$   (c) $\dfrac{1}{10}$   (d) $\dfrac{3}{10}$   **10**(a) $\dfrac{1}{14}$   (b) $\dfrac{1}{7}$   (c) $\dfrac{1}{14}$   (d) $\dfrac{3}{14}$
(e) $\dfrac{1}{6}$   (f) $\dfrac{1}{7}$   **11** $\dfrac{407}{576}$   **12**(a) $\dfrac{37}{64}$   (b) $\dfrac{44}{125}$

### Exercise B

**1**(a) $\tfrac{1}{2}$   (b) $\tfrac{1}{2}$   (c) $\dfrac{5}{36}$   (d) $\tfrac{1}{3}$   (e) $\tfrac{1}{8}$   (f) $\tfrac{1}{3}$   **2**(a) $\tfrac{1}{4}$   (b) $\tfrac{1}{2}$   (c) $\tfrac{1}{2}$
(d) $\tfrac{1}{3}$   (e) $\tfrac{1}{2}$   **3**(a) $\dfrac{2109}{9250}$   (b) $\dfrac{1}{54\,145}$   (c) $\dfrac{33}{108\,290}$   **4** $\dfrac{7}{25}$   **5**(a) $\tfrac{1}{8}$
(b) $\tfrac{5}{9}$   (c) $\dfrac{1}{12}$   **6**(a) $\dfrac{1}{180}$   (b) $\tfrac{1}{3}$   (c) $\tfrac{3}{10}$   **7** $\tfrac{2}{5}$   **8** $\tfrac{1}{4}$   **9**(a) $\dfrac{781}{1024}$

117

(b) $\dfrac{88\,051\,093}{102\,400\,000}$ (c) $\dfrac{14\,348\,907}{102\,400\,000}$ (d) $\dfrac{59\,049}{100\,000}$ **10** 13 **11** $\dfrac{1}{35}$

**12**(a) $\dfrac{3}{10}$ (b) $\dfrac{1}{120}$ (c) $\tfrac{1}{4}$ (d) $\dfrac{7}{12}$ (e) $\dfrac{9}{14}$

## 4.2

**1**(a) yes, 4 (b) yes, $-2$ (c) no (d) yes, $1\tfrac{1}{2}$ (e) yes, $-1\tfrac{1}{4}$
(f) no  **2**(a) 38, $4n-2$, 72, $2n^2$ (b) $-14$, $6-2n$, $-6$, $n(5-n)$
(c) $16\tfrac{1}{2}$, $\tfrac{3}{2}(n+1)$, $40\tfrac{1}{2}$, $\tfrac{1}{4}n(3n+9)$ (d) $-11\tfrac{1}{4}$, $\tfrac{5}{4}(1-n)$, $-18\tfrac{3}{4}$,
$\dfrac{5n}{8}(1-n)$  **3** 1, 4, 7, 10  **4** $a=2, d=1\tfrac{1}{4}$  **5** 32, 29, 26, 23
**6** 10, $-4$  **7** 5  **8** 12  **9** 8, 6, 4, ...  **11**(a) 1.3 (b) 3 (c) 12
**12** 21  **13** 14  **14** 7500  **15** 1900

## 4.3

**1**(a) yes, $1\tfrac{1}{2}$ (b) yes, $\tfrac{3}{4}$ (c) yes, $-3$ (d) no (e) no (f) yes, $\tfrac{1}{3}$
**2**(a) $60\tfrac{3}{4}$, $8(\tfrac{3}{2})^{n-1}$, $166\tfrac{1}{4}$, $16(1.5^n-1)$ (c) $-486$,
$2(-3)^{n-1}$, $-364$, $\tfrac{1}{2}\{1-(-3)^n\}$  **3** 315  **4** $\tfrac{1}{2}$, $17\tfrac{5}{8}$
**5**(a) $n=7$ (b) $n=7$ (c) $n=8$ (d) $n=5$  **6** $r=\tfrac{1}{2}$  **7** $n=5$
**8** 7th term  **9** $n=6$  **10** $a=20$  **11** $a=\pm\tfrac{9}{10}\sqrt{\tfrac{3}{8}}, r=\pm\sqrt{\tfrac{8}{3}}$
**12** 6  **13** $a=3, r=\pm 4$

## 4.4

**1**(a) $1+5x+10x^2+10x^3+5x^4+x^5$
(b) $1+8a+28a^2+56a^3+70a^4+56a^5+28a^6+8a^7+a^8$
(c) $1-7z+21z^2-35z^3+35z^4-21z^5+7z^6-z^7$
**2**(a) $x^5+5x^4y+10x^3y^2+10x^2y^3+5xy^4+y^5$
(b) $c^6-6c^5d+15c^4d^2-20c^3d^3+15c^2d^4-6cd^5+d^6$
(c) $r^7+7r^6s+21r^5s^2+35r^4s^3+35r^3s^4+21r^2s^5+7rs^6+s^7$
**3**(a) 1, $-7$, 21, $-35$ (b) 1, 9, 36, 84, 126
(c) 1, 10, 45, 120, 210, 252  **4**(a) $1+8x+24x^2+32x^3+16x^4$
(b) $1-18y+135y^2-540y^3+1215y^4-1458y^5+729y^6$
(c) $32+240x+720x^2+1080x^3+810x^4+243x^5$
(d) $256-256x+96x^2-16x^3+x^4$  **5**(a) $576x^2$
(b) $17010x^4$ (c) $672x^2y^5$ (d) $5280x^7y^4$  **6**(a) 1140 (b) $-960$
(c) 7 (d) $-\dfrac{28}{9}$ (e) $-15, 120$ (f) $\dfrac{-n(n-1)(n-2)}{3!}a^{n-3}x^3$
**7**(a) $-42$ (b) $-96$  **8** $1+3x+6x^2+7x^3+6x^4+3x^5+x^6$
**9** $1+10x+45x^2+120x^3$  **10** $-3$  **11** 4

## 4.5

**Exercise A**

**1**(a) $1+4+9+16+25$ (b) $1.3+2.4+3.5+4.6$ (c) $\dfrac{1.2}{3}+\dfrac{2.3}{4}+\dfrac{3.4}{5}$

(d) $2+9+28+65$  2(a) $3+12+27+48$  (b) $3.5+4.6+5.7+6.8$
(c) $2+6+12+20$  (d) $1.2.4+2.3.5+3.4.6+4.5.7$  3(a) $\sum_1^7 2r^2$
(b) $\sum_1^9 (r+1)(r+2)$  (c) $\sum \dfrac{(r+1)(r+2)}{r+3}$  (d) $\sum_1^7 27r(r+1)(3r+2)$
(e) $\sum_1^n \dfrac{r^2}{(r+1)}$  4(a) $\dfrac{n(n+1)(4n-1)}{6}$  (b) $\tfrac{1}{2}n(2n^2+11n+3)$
(c) $\tfrac{1}{2}n(n+1)(n^2+n+3)$  (d) $n(n+1)(n^2+3n+1)$.
5 $A=4, B=-6, C=4$; $n(n+1)(n^2+3n+4)$
6 $\dfrac{n}{2(n+2)}$  7 $\dfrac{1-x^n}{(1-x)^2} - \dfrac{nx^n}{1-x}$  8 $\dfrac{n}{12}(n+1)(3n^2+7n+2)$

**Exercise B**

1(a) $\sum_1^7 (2r+1)r^2$  (b) $\sum_1^{10} \dfrac{(2r-1)2r}{(r+2)^2}$  2(a) $\tfrac{1}{3}n(n+1)(n+2)$
(b) $\dfrac{n}{6}(4n^2+21n+29)$  (c) $\dfrac{n}{6}(4n^2+9n-1)$  (d) $n(n+1)^2(n+2)$
3 $A=8, B=-9, C=2$; $n(n+1)(2n^2+7n+7)$
5 $\dfrac{1-(n-1)x^n}{1-x} + \dfrac{x^2(1-x^{n-2})}{(1-x)^2}$  6 $\dfrac{n}{12}(n+1)(3n^2+11n+10)$

## 4.6

1 16, $1\tfrac{1}{2}$  2 £31500  3 1.4 cm  4 £$1.845 \times 10^{17}$  5 7 days, no
6(a) 37.15  (b) 4.355  7 $\dfrac{1-2x+x^{n+2}+n(1-x)}{(1-x)^2}$
8 $a^3+3a^2z+3az^2+z^3$; $\sum a^3 + 3\sum a^2b + 6abc$  10 3, 9, 27, 81 ...
11 4, 12, 20, 28 ...

## 4.7

1(a) $\dfrac{63}{32}$, 2  (b) $\dfrac{728}{243}$, 3  (c) $\dfrac{21}{32}$, $\dfrac{2}{3}$  (d) $\dfrac{3367}{128}$, 32
2 All except (b), (d), (g), (i)  3 125 or $\dfrac{250}{3}$  4 $96+24+6+ \ldots$
5 10  6 11  7 $83\tfrac{1}{3}$  8 $\sec^2\theta$; 4  9 10 or 15  10 8
11 $\cos^2\theta$, $\tfrac{3}{4}$

## 4.8

1(a) $1+\dfrac{1}{2}x-\dfrac{1}{8}x^2+\dfrac{1}{16}x^3 \ldots$  (b) $1+x+x^2+x^3 \ldots$
(c) $1+\dfrac{3}{4}x-\dfrac{3}{32}x^2+\dfrac{5}{128}x^3$  (d) $1+3x+6x^2+10x^3+ \ldots$
2(a) $1-\dfrac{1}{2}x-\dfrac{3}{8}x^2-\dfrac{7}{16}x^3 \ldots$, $-\tfrac{1}{2}<x<\tfrac{1}{2}$

119

(b) $1 - 9x + 54x^2 - 270x^3 \ldots$, $-\frac{1}{3} < x < \frac{1}{3}$
(c) $1 - 4x - 4x^2 + 10\frac{2}{3}x^3 \ldots$, $-\frac{1}{6} < x < \frac{1}{6}$
(d) $1 + 2x + 2\frac{1}{2}x^2 + 2\frac{1}{2}x^3 + \ldots$, $-2 < x < 2$

3(a) $1 + 2x + 6x^2$, $-\frac{1}{4} < x < \frac{1}{4}$  (b) $2 - \frac{3}{4}x - \frac{9}{64}x^2$, $-\frac{4}{3} < x < \frac{4}{3}$

(c) $4 - \frac{5}{3}x - \frac{25}{144}x^2$, $-\frac{8}{5} < x < \frac{8}{5}$  (d) $\frac{1}{81}(3 + 12x + 32x^2)$, $-\frac{3}{4} < x < \frac{3}{4}$

(e) $1 - x + 2x^2$, $|x| < \frac{1}{3}$   4 $\frac{1}{4}(3x^2 + 8x^3)$   5 $2x - \frac{5}{2}x^2 + \frac{25}{6}x^3$,
$|3x| < 1$   6(a) $1 - 3x^2 - x^3$, $|2x| < 1$   (b) $1 - 2x + 2x^2$, $|x| < 1$
(c) $1 + 3x + 7\frac{1}{2}x^2$, $|4x| < 1$   (d) $1 - 4x + 5x^2$, $|4x| < 1$   7 $1 + x^2 + x^4 + x^6$
8 $1 + x + x^2 + x^3 + \ldots$, $1 + x + x^2$

## 4.9

1(a) $1 + 2 + \frac{2^2}{2!} + \frac{2^3}{3!} + \frac{2^4}{4!}$   (b) $1 + \frac{1}{2} + \frac{1}{2!2^2} + \frac{1}{3!2^3} + \frac{1}{4!2^4}$

(c) $1 - \frac{1}{3} + \frac{1}{2!3^2} - \frac{1}{3!3^3} + \frac{1}{4!3^4}$   2(a) $1 - x + \frac{x^2}{2!} - \frac{x^3}{3!} + \frac{x^4}{4!}$

(b) $1 - 2x + \frac{4x^2}{2!} - \frac{8x^3}{3!} + \frac{16x^4}{4!}$   (c) $1 + \frac{1}{2}x + \frac{x^2}{2!2^2} + \frac{x^3}{3!2^3} + \frac{x^4}{4!2^4}$

(d) $1 - \frac{1}{x} + \frac{1}{2!x^2} - \frac{1}{3!x^3} + \frac{1}{4!x^4}$   (e) $1 - x^2 + \frac{x^4}{2!} - \frac{x^6}{3!} + \frac{x^8}{4!}$

(f) $1 + \frac{1}{x^2} + \frac{1}{2!x^4} + \frac{1}{3!x^6} + \frac{1}{4!x^8}$   3(a) $e^5$   (b) $e^{1/2}$   (c) $e^{0.1}$   (d) $e^{-4}$

(e) $e^{-0.3}$   4(a) $1 + \frac{3^2}{2!} + \frac{3^4}{4!} + \ldots$   (b) $4 + \frac{4^3}{3!} + \frac{4^5}{5!} + \ldots$

5(a) $1 + 2x + \frac{3}{2}x^2 + \frac{2}{3}x^3 + \ldots$   (b) $1 - 3x + \frac{5}{2}x^2 - \frac{7}{6}x^3 + \ldots$

(c) $2 - 4x + 5x^2 - \frac{14}{3}x^3 + \ldots$   (d) $2x + \frac{2x^2}{2!} + \frac{2x^4}{4!} + \frac{2x^6}{6!} + \ldots$

(e) $e(1 + x^2 + \frac{x^4}{2!} + \frac{x^6}{3!} + \ldots)$   (f) $1 + \frac{9}{2}x + \frac{17}{8}x^2 + \frac{25}{48}x^3$

(g) $1 + \frac{x^2}{2} + \frac{x^4}{24} + \frac{x^6}{720}$

6 Successive terms are 1, 4, 8, 10.667, 10.667, 8.533, 5.689, 3.251, 1.625, 0.722, 0.289, 0.105, 0.035, 0.011.   7(a) $\frac{1}{2} - \frac{1}{2 \cdot 2^2} + \frac{1}{3 \cdot 2^3} - \frac{1}{4 \cdot 2^4} + \ldots$

(b) $0.2 - \frac{0.04}{2} + \frac{0.008}{3} - \frac{0.0016}{4}$   (c) $-0.1 - \frac{0.01}{2} - \frac{0.001}{3} - \frac{0.0001}{4}$

(d) $\frac{1}{4} - \frac{1}{32} + \frac{1}{192} - \frac{1}{1024} + \ldots$   (e) $-\frac{1}{8} - \frac{1}{128} - \frac{1}{1536} - \frac{1}{16384}$

8(a) 0.009 550   (b) $-0.020\,203$

**9** $e^x \cdot e^y = 1 + (x+y) + \dfrac{(x+y)^2}{2!} + \dfrac{(x+y)^3}{3!} + \ldots$

**10**(a) $2x - 2x^2 + \dfrac{8}{3}x^3$, $-\tfrac{1}{2} < x \leqslant \tfrac{1}{2}$  (b) $-\dfrac{1}{2}x - \dfrac{1}{8}x^2 - \dfrac{1}{24}x^3$, $-2 \leqslant x < 2$

(c) $x^2 - \dfrac{x^4}{2} + \dfrac{x^6}{3}$, $0 \leqslant x \leqslant 1$  (d) $-\sqrt{x} - \dfrac{x}{2} - \dfrac{x\sqrt{x}}{3}$, $0 < x < 1$

(e) $-2(x + \tfrac{1}{2}x^2 + \tfrac{1}{3}x^3)$, $-1 \leqslant x < 1$  (f) $3x - \dfrac{5}{2}x^2 + \dfrac{9}{3}x^3$, $-\tfrac{1}{3} < x \leqslant \tfrac{1}{3}$

(g) $3x - \dfrac{3}{2}x^2 + 3x^3$, $-\tfrac{1}{2} < x \leqslant \tfrac{1}{2}$  (h) $-5x - \dfrac{13}{2}x^2 - \dfrac{35}{3}x^3$, $-\tfrac{1}{3} < x \leqslant \tfrac{1}{3}$

## 4.10

**1**(a) $2x - \dfrac{8x^3}{3!} + \dfrac{32x^5}{5!}$  (b) $1 - \dfrac{x^4}{2!} + \dfrac{x^8}{4!}$  (c) $\dfrac{x}{3} - \dfrac{x^3}{162} + \dfrac{x^5}{29\,160}$

**3**(a) $\dfrac{2x^2}{2!} - \dfrac{8x^4}{4!} + \dfrac{32x^6}{6!}$  (b) $x + \dfrac{x^3}{3} + \dfrac{2x^5}{15}$  (c) $1 + x + \tfrac{1}{2}x^2$

(d) $1 + x - \dfrac{2x^3}{3!}$  (e) $1 + x\log 2 + \dfrac{(x\log 2)^2}{2!}$  (f) $x - \tfrac{1}{2}x^2 + \tfrac{1}{6}x^3$

**4**(a) $x + \dfrac{5x^3}{6} - \dfrac{19x^5}{120} + \dfrac{41x^7}{5040}$, valid for all $x$

(b) $x + x^2 + \tfrac{1}{3}x^3 - \tfrac{1}{30}x^5$, valid for all $x$  (c) $-2x^2 - x^3 + \tfrac{2}{3}x^4 - \tfrac{1}{6}x^5$,

$-1 \leqslant x < 1$  (d) $x + \tfrac{1}{2}x^2 - \tfrac{2}{3}x^3 + \tfrac{1}{4}x^4$, $-1 \leqslant x < 1$  **5**(a) $x \sin x$

(b) $\sin x + \cos x$  (c) $\cos x - (1-x)$  **6**(a) $0.09983$  (b) $0.5000$

(c) $0.1736$  **7** $-\dfrac{3}{8} \cdot \dfrac{(x-2)^4}{4!} + \dfrac{3}{4} \cdot \dfrac{(x-2)^5}{5!}$  **8** $x - \dfrac{x^3}{3} + \dfrac{x^5}{15} - \dfrac{x^7}{105}$

**9** $\sin\theta + x\cos\theta - \dfrac{x^2}{2!}\sin\theta - \dfrac{x^3}{3!}\cos\theta$

## 4.11

**2** $1 + \dfrac{1}{2}x - \dfrac{1}{8}x^2 + \dfrac{1}{16}x^3$, $1 - 2x + 3x^2 - 4x^3$, $4.009\,988$,

$0.990\,075$  **3** $1 - \dfrac{1}{2}x - \dfrac{1}{8}x^2 - \dfrac{1}{16}x^3$, $3.316\,625$  **4** $a = 3$, $n = -4$,

$p = -540$; $(1 + \tfrac{3}{2}x)^{-4}$  **5** $16$, $a = 2$  **7** $1 - \dfrac{1}{4}x - \dfrac{3}{32}x^2 - \dfrac{7}{128}x^3$,

$1.49535$.  **8** $1 + \dfrac{1}{2}x - \dfrac{1}{8}x^2 + \dfrac{1}{16}x^3$, $1.0247$, $2.47\%$  **9** $1 - 6x$

**10** $7.1414$  **11** $a = 4$, $b = 2$; $-5\tfrac{1}{3}$  **12** $y > 1$; $0.6931$

**13** $x + \tfrac{1}{2}x^2 + \tfrac{1}{3}x^3$; no; no $x^4$ term; $0.105\,334$  **14** $0.041393$

**15** $1 - \dfrac{1}{2}x - \dfrac{1}{8}x^2 - \dfrac{1}{16}x^3$; $1 - 2x + 3x^2 - 4x^3$

**16** $x^2 + x^3 + \dfrac{1}{2}x^4 + \dfrac{1}{6}x^5 - \dfrac{1}{8}x^6$; $2x + 3x^2 + 2x^3 + \dfrac{5}{6}x^4 - \dfrac{3}{4}x^5$

**17** $\frac{1}{4}\pi + \frac{1}{2}x - \frac{1}{4}x^2 + \frac{1}{12}x^3$   **18** $a = 1, b = \pm 2$

## 5.2

### Exercise A

**1**(a) $-\frac{1}{\sqrt{2}}$   (b) 0   (c) $-1$   (d) $\frac{\sqrt{3}}{2}$   (e) $\frac{\sqrt{3}}{2}$   (f) $-\frac{2}{\sqrt{3}}$
(g) $-\sqrt{3}$   (h) 1   (i) $\frac{\sqrt{3}}{2}$   (j) $-\sqrt{2}$   **2**(a) $\frac{1}{3}\pi, \frac{2}{3}\pi$   (b) $\frac{2}{3}\pi, \frac{4}{3}\pi$
(c) $\frac{3}{4}\pi, \frac{7}{4}\pi$   (d) $\frac{1}{4}\pi, \frac{7}{4}\pi$   (e) $\frac{7}{6}\pi, \frac{11}{6}\pi$   (f) $\frac{1}{6}\pi, \frac{7}{6}\pi$   (g) $\frac{5}{4}\pi, \frac{7}{4}\pi$
(h) $\frac{1}{4}\pi, \frac{7}{4}\pi$   **3**(a) $-0.5000$   (b) 0.8660   (c) $-11.4300$   (d) $-0.5095$
(e) $-0.9848$   (f) $-3.2361$   (g) 0.7660   (h) $-1.2521$   (i) 0.4540
(j) $-0.3420$   (k) $-0.9397$   (l) 19.081   **4**(a) 0.6109   (b) 1.9548
(c) 3.3685   (d) 6.1959   (e) 12.4268   (f) 14.6608   **5**(a) 120°
(b) 112° 30′   (c) 300°   (d) 57° 18′   (e) 68° 45′   (f) 85° 57′
(g) 42° 58′   (h) 200° 33′   **8**(a) $\frac{1}{\sqrt{2}}$   (b) 0   (c) $-0.7265$

(d) $-0.3249$   (e) 1.0824   (f) $\sqrt{2}$   (g) 0.9316   (h) $-0.9093$
(i) $-0.9668$   (j) $-1.9399$   **9**(a) $-330°, -210°, 30°, 150°$
(b) $\pm 103° 20′, \pm 256° 40′$   (c) $-284° 54′, -4° 54′,$
75° 6′, 255° 6′   (d) $-156° 25′, -23° 35′, 203° 35′, 336° 25′$
(e) $-318° 36′, -41° 24′, 41° 24′, 318° 36′$   (f) $-337° 36′,$
$-202° 24′, 22° 24′, 157° 36′$   (g) $-254° 43′, -74° 43′, 105° 17′,$
285° 17′   (h) $-304° 58′, -55° 2′, 55° 2′, 304° 58′$   (i) $-144° 24′,$
$-35° 36′, 215° 36′, 324° 24′$   (j) $-333° 37′, -206° 23′,$
26° 23′, 153° 37′   **10**(a) $-80° 52′, 20° 52′$   (b) 66° 9′,
$-146° 9′$   (c) $-43° 17′, 43° 17′$   (d) 62° 31′, 117° 29′   (e) $-52° 36′,$
$-142° 36′, 37° 24′, 127° 24′$   (f) 33° 42′, $-146° 18′$   (g) 37° 10′,
$-37° 10′, 142° 50′, -142° 50′$   (h) 59° 35′, $-59° 35′, 120° 26′,$
$-120° 26′$   (i) $\pm 19° 11′, \pm 40° 49′, \pm 139° 11′, \pm 79° 11′,$
$\pm 100° 49′, \pm 199° 11′$   (j) 0° 33′, 119° 27′, $-60° 33′, -179° 27′$
(k) $-7° 22′, -42° 38′, 47° 42′, 82° 38′, 172° 38′, 137° 23′$
**11**(a) 0°, 180°, 90°, 270°, 360°   (b) 120°, 240°, 48° 11′, 311° 49′   (c) 90°,
270°   (d) 199° 28′, 340° 32′   (e) 45°, 225°, 116° 34′, 296° 34′   (f) 78° 27′,
281° 33′   (g) 90°, 270°   (h) 56° 18′, 236° 18′   (i) 0°, 180°, 360°, 41° 24′,
318° 36′   (j) 199° 28′, 340° 32′   (k) 28° 8′, 151° 52′, 208° 8′, 331° 52′
(l) 30°, 150°, 194° 29′, 345° 31′   (m) 0°, 180°, 360°, 60°, 120°, 240°, 300°
(n) 19° 28′, 160° 32′   (o) 72° 23′, 287° 37′   **12**(a) $\cos\theta$   (b) $-\cos\theta$
(c) $-\tan\theta$   (d) $-\cot\theta$   (e) $\tan\theta$   (f) $-\tan\theta$   (g) $-\cos\theta$
(h) $-\cot\theta$   (i) $\operatorname{cosec}\theta$   (j) $-\operatorname{cosec}\theta$   (k) $-\tan\theta$   (l) $-\tan\theta$

### Exercise B

**3**(a) $\frac{1}{4}\pi, \frac{7}{4}\pi$   (b) $\frac{4}{3}\pi, \frac{5}{3}\pi$   (c) $\frac{5}{6}\pi, \frac{11}{6}\pi$   (d) $\frac{3}{4}\pi, \frac{5}{4}\pi$   (e) $\frac{1}{4}\pi, \frac{3}{4}\pi$

(f) $\frac{1}{3}\pi$, $\frac{4}{3}\pi$  (g) $\frac{5}{4}\pi$, $\frac{7}{4}\pi$  4(a) $\frac{1}{3}\pi$, $\frac{5}{3}\pi$, $-\frac{1}{3}\pi$, $-\frac{5}{3}\pi$  (b) 0.32, $\pi - 0.32$, $0.32 - 2\pi$, $-(\pi + 0.32)$  (c) 1.08, $\pi + 1.08$, $1.08 - 2\pi$, $1.08 - \pi$
(d) 0.443, $\pi - 0.443$, $0.443 - 2\pi$, $-(\pi + 0.443)$  (e) 0.775, $2\pi - 0.775$, $-0.775$, $0.775 - 2\pi$  (f) 1.074, $\pi - 1.074$, $1.074 - 2\pi$, $-(\pi + 1.074)$
(g) $\pi - 1.242$, $2\pi - 1.242$, $-1.242$, $-(1.242 + \pi)$  (h) 1.288, $2\pi - 1.288$, $-1.288$, $-2\pi + 1.288$  (i) $\pi + 0.433$, $2\pi - 0.433$, $-0.433$, $-\pi + 0.433$
(j) 0.731, $\pi - 0.731$, $-2\pi + 0.731$, $-(\pi + 0.731)$  5(a) $-101°13'$, $-18°47'$  (b) $41°25'$, $178°35'$  (c) $74°4'$, $-105°56'$
(d) $35°22'$, $144°38'$  (e) $-13°55'$, $-103°55'$, $76°5'$, $166°5'$
(f) $\pm 22°21'$, $\pm 97°39'$, $\pm 142°21'$  (g) $\pm 15°56'$, $\pm 74°4'$, $\pm 105°56'$, $\pm 164°4'$  (h) $52°21'$, $7°39'$, $-127°39'$, $-172°21'$
(i) $26°40'$, $66°40'$, $-33°20'$, $6°40'$, $86°40'$, $-53°20'$, $126°40'$, $-93°20'$, $146°40'$, $-113°20'$, $173°20'$, $-153°20'$,  (j) $\pm 20°2'$, $\pm 51°58'$, $\pm 92°2'$, $\pm 123°58'$, $\pm 164°2'$  (k) $-166°40'$, $-146°40'$, $-106°40'$, $-86°40'$, $-46°40'$, $-26°40'$, $13°20'$, $33°20'$, $73°20'$, $93°20'$, $133°20'$, $153°20'$  6(a) $90°$, $270°$, $180°$  (b) $30°$, $150°$, $270°$
(c) $0°$, $180°$, $360°$, $146°19'$, $326°19'$  (d) $60°$, $300°$, $0°$, $360°$
(e) $45°$, $225°$, $104°2'$, $284°2'$  (f) $18°26'$, $198°26'$
(g) $90°$, $270°$, $41°49'$, $138°11'$  (h) $65°54'$, $114°6'$, $245°54'$, $294°6'$
(i) $0°$, $360°$  (j) $39°14'$, $140°46'$, $219°14'$, $320°46'$
(k) $27°36'$, $152°24'$, $239°42'$, $300°18'$  (l) $90°$, $270°$
(m) $16°36'$, $163°27'$  (n) $90°$, $270°$, $60°$, $300°$  7(a) $\sin\theta$
(b) $-\sin\theta$  (c) $\cot\theta$  (d) $\cot\theta$  (e) $\sec\theta$  (f) $-\sin\theta$  (g) $\sec\theta$
(h) $-\tan\theta$  (i) $-\cot\theta$  (j) $-\tan^2\theta$

## 5.3

**Exercise A**

3(a) $x^2 + y^2 = a^2$  (b) $\dfrac{y^2}{a^2} - \dfrac{x^2}{b^2} = 1$  (c) $x\sqrt{a^2 - y^2} = y$
(d) $x^2 + y^2 = 2$  (e) $x^2 - y^2 = 1$  (f) $1 + y^2 = x^2 y^4$  (g) $1 + (x-a)^2 = y^2$
4(a) $\theta = 90°$, 4; $\theta = 270°$, $-2$  (b) $\theta = 180°$, 6; $\theta = 360°$, 2
(c) $\theta = 180°$, $\infty$; $\theta = 360°$, $\frac{1}{2}$  (d) $\theta = 90°$, 1; $\theta = 45°$, 0
(e) $\theta = 45°$, $\frac{1}{2}$; $\theta = 135°$, $\frac{1}{8}$  (f) $\theta = 30°$, 4; $\theta = 210°$, $-4$  5(a) $\dfrac{3}{4}$
(b) $\dfrac{12}{13}$  (c) $\dfrac{15}{17}$  (d) $\dfrac{5}{4}$  6(a) $180°$  (b) $315°$  (c) $240°$  (d) $240°$

**Exercise B**

3(a) $\dfrac{x^2}{9} + \dfrac{y^2}{16} = 1$  (b) $x^2 + y^2 = 13$  (c) $\dfrac{x^2}{4} - \dfrac{1}{y^2} = 1$  (d) $xy = 1$
(e) $\dfrac{1}{y^2} - (x+y)^2 = 1$  (f) $\left(1 + \dfrac{x}{y}\right)^2 + \left(\dfrac{y}{x}\right)^2 = 1$  4(a) (i) $\theta = 270°$, 3
(ii) $\theta = 90°$, 1  (b) (i) $\theta = 360°$, 2 (ii) $\theta = 180°$, $-4$  (c) (i) $\theta = 90°$, 2
(ii) $\theta = 180°$, $\frac{2}{3}$  (d) (i) $\theta = 360°$, 3 (ii) $\theta = 720°$, 0  (e) (i) $\theta = 120°$, $\frac{1}{2}$
(ii) $\theta = 60°$, $\frac{1}{4}$  5(a) $\dfrac{\sqrt{15}}{4}$  (b) $\dfrac{2}{\sqrt{5}}, \dfrac{3}{\sqrt{5}}$  (c) $\dfrac{1}{\sqrt{24}}, \dfrac{5}{\sqrt{24}}$

(d) $\dfrac{2}{\sqrt{53}}, \dfrac{\sqrt{53}}{7}$   6(a) 120°   (b) 225°   (c) 300°   (d) 150°

## 6.2

**Exercise A**

1(a) $\dfrac{\sqrt{3}-1}{2\sqrt{2}}$   (b) $-\dfrac{(1+\sqrt{3})}{2\sqrt{2}}$   (c) $-(2+\sqrt{3})$   (d) $\dfrac{2\sqrt{2}}{1-\sqrt{3}}$

(e) $\dfrac{\sqrt{3}+1}{2\sqrt{2}}$   (f) $-(2+\sqrt{3})$   2(a) $\dfrac{\sqrt{3}}{2}\sin x + \tfrac{1}{2}\cos x$

(b) $\dfrac{1}{\sqrt{2}}(\cos x + \sin x)$   (c) $-\sin x$   (d) $\cot x$   (e) $\cot x$

(f) $-(\sin x + \tfrac{1}{2}\cos x)$   3(a) $\dfrac{1}{\sqrt{2}}$   (b) $\dfrac{1}{2}$   5(a) $\dfrac{56}{33}$   (b) $\dfrac{63}{65}$   (c) $\dfrac{119}{120}$

6(a) $\dfrac{56}{65}$   (b) $\dfrac{63}{65}$   (c) $-\dfrac{56}{33}$

7(a) $\sin 9°$   (b) $\tan 45° = 1$   (c) $\tan 30° = \dfrac{1}{\sqrt{3}}$   (d) $\sin(60° - x)$

(e) $\cos(45° - x)$   (f) $\sin 30° = \tfrac{1}{2}$   (g) $\cos 60° = \tfrac{1}{2}$   (h) 1

(i) $\cot 30° = \sqrt{3}$   8(a) $x = 0°, 180°$   (b) 105° 57′   (c) 3° 31′

(d) 90°, 36° 52′   (e) 23° 53′, 83° 53′, 143° 53′   10 $\dfrac{2-\sqrt{2}}{\sqrt{6}-2}$

11 $\dfrac{7}{6}$   12 $\dfrac{120}{169}, -\dfrac{119}{169}$   13 $5\sqrt{3}-8$   14(a) $\dfrac{1}{\sqrt{2}}$   (b) $\dfrac{\sqrt{3}}{2}$

16(a) 52° 30′, 232° 30′   (b) 16° 20′, 196° 20′   (c) 40°, 220°

**Exercise B**

1(a) $\dfrac{1}{2\sqrt{2}}(\sqrt{3}+1)$   (b) $\dfrac{\sqrt{3}-1}{\sqrt{3}+1}$   (c) $\dfrac{1}{2\sqrt{2}}(\sqrt{3}+1)$   (d) $\dfrac{\sqrt{3}-1}{\sqrt{3}+1}$

(e) $\dfrac{1}{2\sqrt{2}}(\sqrt{3}+1)$   (f) $\dfrac{\sqrt{3}+1}{1-\sqrt{3}}$   (g) $\dfrac{1}{2}\sqrt{\dfrac{1}{\sqrt{2}}(\sqrt{3}+2\sqrt{2}+1)}$

2(a) $\dfrac{77}{85}$   (b) $\dfrac{84}{85}$   (c) $\dfrac{240}{161}$   3 $\dfrac{2(1+\tan^2 A)}{1-\tan^2 A}$   4(a) $\sin(x+30°)$

(b) $2\sin(x-60°)$   (c) $\tan(x+45°)$   (d) $\dfrac{1}{2}$   (e) $\cot(60°-x)$

5 $\dfrac{7}{6}$   6 1   7(a) $\cos(x+60°)$   (b) $\sqrt{2}\cos(x-45°)$   (c) $\tan(60°-x)$

(d) $\dfrac{\sqrt{3}}{2}$   (e) $\tfrac{1}{4}\sin x$   (f) 1   8 $\dfrac{336}{625}$   $\dfrac{10296}{15625}$   $\dfrac{336}{527}$   $\dfrac{11753}{15625}$

11(a) 120°, −60°   (b) 45°, −135°   (c) 5° 10′, −174° 50′, −65° 15′,

114° 45'   **16**(a) $\dfrac{23}{25}, \dfrac{71}{125}$   (b) $\pm\dfrac{\sqrt{15}}{4}, \pm\dfrac{\sqrt{3}}{2\sqrt{2}}, \pm\dfrac{\sqrt{5}}{2\sqrt{2}}$

(c) $\sqrt{\dfrac{\sqrt{2}-1}{2\sqrt{2}}}, \sqrt{\dfrac{\sqrt{2}+1}{2\sqrt{2}}}$   **17**(a) 60°, 240°

(b) 63° 26', 243° 26'   (c) 126° 52'   (d) 30°, 150°
(e) 153° 26', 333° 26'   (f) $22\tfrac{1}{2}°$, 90°, $112\tfrac{1}{2}°$, $202\tfrac{1}{2}°$, $292\tfrac{1}{2}°$, 270°
(g) no solutions   (h) 33° 46', 146° 14', 213° 46', 326° 14'   **18**(a) $\dfrac{4}{3}$

(b) $\pm\dfrac{1}{\sqrt{3}}, \pm\sqrt{\dfrac{2}{3}}$   (c) $-\dfrac{120}{169}, \dfrac{119}{169}$   (d) $\sqrt{2}-1$
**19**(a) 51° 20', 128° 40'   (b) 0°, 180°, 360°, 70° 32', 289° 28'   (c) 0°, 180°, 360°, 40° 39', 139° 21'   (d) 0°, 180°, 360°, 39° 14' 129° 24', 230° 46', 309° 24'   (e) 0°, 180°, 360°, 79° 10', 100° 50', 219° 14', 320° 46'

## 6.3

**Exercise A**
**1**(a) 2 cos 45° cos 15°   (b) 2 cos 30° sin 10°   (c) 2 sin 55° sin 15°
(d) 2 cos (x + 15°) sin 15°   (e) $2 \sin (x - \tfrac{1}{2}h) \sin \tfrac{1}{2}h$   (f) 2 sin x cos h
**2**(a) sin 90° + sin 30°   (b) $\tfrac{1}{2}(\cos 4x - \cos 10x)$   (c) cos 2A + cos 2B
(d) sin 2y + sin 2(x − z)   (e) $\tfrac{1}{2}(\cos 6\pi + \cos x)$   **8**(a) 0°, 180°, 360°, 120°, 240°   (b) 0°, 51° 26', 102° 51', 154° 17', 205° 43', 257° 9', 308° 34', 360°   (c) 75°, 255°   (d) 33° 20', 93° 20', 153° 20', 213° 20', 273° 20', 333° 20'   (e) 30° 31', 149° 29'
(f) 21° 55', 162° 5'   (g) 67° 30', 157° 30', 247° 30', 337° 30', 135°, 315°   (h) 100° 1', 164° 58', 280° 1', 244° 58'
(i) 35° 15', 57° 15', 125° 15', 147° 15', 215° 15', 237° 15', 305° 15', 327° 15'   (j) 22° 2', 82° 2', 142° 2', 202° 2', 262° 2', 322° 2', 36° 58', 96° 58', 56° 58', 216° 58', 276° 58', 336° 58', 105° 12', 165° 12', 225° 12', 285° 12', 345° 12'   (k) 0°, 60°, 120°, 180°, 240°, 300°, 360°

**Exercise B**
**1**(a) 2 sin 3x cos x   (b) 2 cos 6x cos 2x   (c) 2 cos 6x sin x
(d) $2 \sin \tfrac{3}{2}x \cos x$   (e) $2 \sin (30° + \tfrac{1}{2}x) \cos (30° - \tfrac{1}{2}x)$
(f) 2 cos 45° cos (45° − x)   (g) 2 sin (45° + x) cos (45° − x)
(h) 2 sin x cos 60°   (i) −2 sin (x − 30°) sin 46°   **2**(a) cos 60° + cos 20°
(b) $\tfrac{1}{2}(\sin 6x - \sin 2\alpha)$   (c) $\tfrac{1}{2}(\cos 2x - \cos 6x)$   (d) sin x + sin 30°
(e) $\cos \tfrac{1}{2}\pi + \cos \tfrac{1}{4}\pi$   (f) $\tfrac{1}{2}(\sin 3y - \sin \tfrac{1}{3}y)$   **8**(a) ±52° 10'
(b) 19° 43', 152° 17'   (c) 90°, −30°, −150°   (d) no solutions
(e) 9° 31', 49° 9', 109° 9', −50° 29', −110° 29', −170° 29', 10° 51', −70° 51', −130° 51'   (f) 10° 24', 100° 24', −44° 24', −134° 24'   (g) ±30°, ±90°, ±150°, ±60°, ±120°   (h) 0°, ±45°, ±90°, ±135°, ±60°, ±120°
(i) ±30°, ±90°, ±150°   (j) 0°, ±90°, ±180°, ±51° 26', 102° 51', ±154° 17'   (k) 0°

## 6.4

**2**(a) 0.74  (b) 0.6  (c) 0.3  (d) 0.51  **4**(a) $2\theta$  (b) $1-\frac{9}{2}\theta^2$  (c) $\frac{5}{8\theta}$
(d) $\frac{1}{2}$  (e) 16  (f) $-\theta$  (g) $\theta$  (h) $\frac{1}{(\cos h - \theta \sin h)\cosh h}$
(i) 1  **6**(a) 0.20  (b) 1.24  (c) 0.88  (d) 0.97  **7**(a) $-2\sin h$
(b) $\frac{9}{16}$  (c) 1  (d) $1-8\theta^2$  (e) 1  (f) $2\theta+1$

## 6.5

**2**(a) 30°  (b) 31° 36′  (c) $-0.1736$  **3**(a) 30°  (b) 40° 54′
(c) 16° 38′  (d) 12° 57′  (e) 1  **6** 0.9899
**8**(a) $5\cos(\theta - 53° 8')$  (b) $\sqrt{13}\cos(\theta + 56° 18')$
**9**(a) $13\sin(2\theta + 67° 23')$  (b) $\sqrt{45}\sin(2\theta - 26° 34')$
(c) $-\sqrt{73}\sin(2\theta - 69° 27')$  **10**(a) max, 5, when $\theta = 53° 8'$;
min, $-5$, when $\theta = 233° 8'$  (b) max, 2, when $\theta = 60°$; min,
$-2$, when $\theta = 240°$  (c) max, 10, when $\theta = 71° 34'$;
min, 0, when $\theta = 161° 34'$  (d) max, $\sqrt{2}$, when $\theta = 157\frac{1}{2}°$; min,
$-\sqrt{2}$, when $\theta = 67\frac{1}{2}°$  (e) max, 13, when $\theta = 67° 32'$;
min, $-13$, when $\theta = 7° 32'$  **11**(a) 306° 52′  (b) 130° 12′, 342° 26′
(c) 12° 3′, 124° 21′  (d) 124° 49′, 325° 11′  (e) 27° 37′, 180°  (f) no solutions  **12**(a) 0°, 180°, 360°, 63° 26′, 243° 26′  (b) 95° 27′, 275° 27′, 25° 31′, 205° 31′  (c) 107° 59′, 287° 59′, 4° 38′, 184° 38′  (d) 96° 39′, 276° 39′, 30° 13′, 210° 13′  **13**(a) 0°, 126° 52′  (b) $-163° 44'$, 90°
(c) 97° 50′, $-26° 45'$  (d) $-136° 5'$, 9° 13′  **14**(a) $360n° \pm 36°$
or $180n°$  (b) $\frac{1}{2}\pi n - \frac{1}{24}\pi$  (c) $360n° \pm 144°$
(d) $180n°$ or $180n° + (-1)^n 52° 14'$ or $180n° - (-1)^n 52° 14'$
(e) $360n° \pm 66° 25' + 36° 52'$  (f) $60n°$ or $90n° \pm 30°$  (g) $120n° \pm 30°$
or $360n° \pm 60°$  (h) $72n° - 8°$ or $360n° \pm 180° + 80°$  (i) $360n° \pm 90°$
or $72n°$ or $360n° \pm 180°$  (j) $180n° \pm (-1)^n 30°$ or $180n°$
or $360n° \pm 90°$  (k) $\frac{n}{6}\pi - \frac{1}{18}\pi$

## 7.2

**3**(a) $b = 9.132$ cm, $c = 14.82$ cm, $C = 88°$, area $= 54.76$ cm$^2$
(b) $b = 59.10$ m, $c = 49.20$, $C = 43°$, area $= 302.3$ m$^2$
(c) $C = 25° 58'$, $A = 107° 50'$, $a = 16.09$ mm, area $= 42.97$ mm$^2$
(d) $b = 11.28$ cm, $c = 22.81$ cm, $A = 40° 32'$, area $= 83.61$ cm$^2$
(e) $b = 4.108$ cm, $c = 18.8$ cm, $A = 65° 30'$, area $= 35.14$ cm$^2$
(f) $A = 31° 59'$, $B = 42° 23'$, $C = 105° 38'$
(g) $A = 57° 33'$, $B = 75° 10'$, $C = 47° 17'$, area $= 16.17$ cm$^2$
(h) $B = 21° 21'$, $A = 140° 30'$, $a = 20.96$ m, area $= 41.98$ m$^2$ or
$B = 158° 39'$, $A = 30° 12'$, $a = 1.84$ m, area $= 3.68$ m$^2$
(i) $C = 72° 6'$, $b = 10.34$ cm, $c = 15.15$ cm, area $= 72.31$ cm$^2$

**4** 0.2490 cm², 4.1465 cm   **5** $\left(\dfrac{\pi\theta}{180} - \sin\theta\right)\dfrac{r^2}{2}$ cm²,

$\dfrac{\pi\theta}{180}r + 2r\sin\tfrac{1}{2}\theta$ cm   **6** 0.08a²   **7** 0.1133a²   **8** 0.0007a²,

0.2003a   **10** $\left(\pi - \dfrac{3\sqrt{3}}{2}\right)a^2$   **11** $\dfrac{9}{4}(\tfrac{1}{2}\pi + 3)$ units

**12** 18.79 cm²   **13** 11.43:1   **15** 9.641 cm²   **16** $\dfrac{\theta + \sin\theta - \sin 2\theta}{(2\pi - \theta + \sin 2\theta - \sin\theta)}$

## 7.4

**1**(a) 35° 16′   (b) 26° 34′   (c) $\cos^{-1}\sqrt{\dfrac{2}{3}}$   **2** $\cos^{-1}\dfrac{1}{3}$

**3** $\cos^{-1}\dfrac{9}{10}$, $\cos^{-1}\dfrac{4}{\sqrt{19}}$   **4**(a) $\tan^{-1}\dfrac{14}{5}$   (b) $\cos^{-1}\dfrac{171}{221}$

(c) $\cos^{-1}\dfrac{-25}{221}$   **6** $\cos^{-1}\dfrac{3\sqrt{5}}{2}$   **7** $\dfrac{4}{3\sqrt{5}}$

**10** $\cos^{-1}\dfrac{2k^2 - 3k + 2}{\sqrt{2k^2 - 2k + 1}\sqrt{3k^2 - 4k + 4}}$, $\cos^{-1}\dfrac{1-k}{\sqrt{1+k^2}\sqrt{(k-1)^2 + k^2}}$

## 8.2

### Exercise A

**1**(a) $\tan\theta = 2$   (b) $r = 4a\,\text{cosec}\,\theta\cot\theta$   (c) $r = \dfrac{5}{\sin\theta + 3\cos\theta}$

(d) $r^2 = 2c^2\,\text{cosec}\,2\theta$   (e) $r^2 = \dfrac{\sin 2\theta}{2 - \sin^2 2\theta}$   (f) $r\cos\theta = 0$

(g) $r^2 - 2r(a\cos\theta + b\sin\theta) = 1 - a^2 - b^2$   **2**(a) $x^2 + y^2 = ay$
(b) $x^2 + y^2 = k^2$   (c) $y = x\tan k$   (d) $(x^2 + y^2)^{3/2} = ax^2$
(e) $(x^2 + y^2)^2 = 2axy$   (f) $x^2 + y^2 = a(\sqrt{x^2 + y^2} - x)$
(g) $3 - 2x = \sqrt{x^2 + y^2}$   **3**(a) $r = c$   (b) $\theta = \tan^{-1} m$
(c) $r = 2a\cos\theta$   **6**(a) $y^2 = 4(x - 1)$   (b) $x^2 + y^2 = 4$
(c) $x^2 - y^2 = 4$   (d) $x^3 y^2 = 1$   (e) $x = 4y^3 - 3y$   **7**(a) $(4at^2, 4at)$
(b) $\left(\dfrac{3t}{\sqrt{1+t^2}}, \dfrac{3}{\sqrt{1+t^2}}\right)$   (c) $\left(\dfrac{t}{t-1}, \dfrac{1}{t-1}\right)$

### Exercise B

**1**(a) $r^2 + 3a^2 = 4ar\cos\theta$   (b) $\sqrt{2}r\sin(45° - \theta) = a$   (c) $r = 2a\sin\theta$
**4**(a) $4xy + x + y = 2$   (b) $y(x - 1)^2 = 1$   (c) $y = 1 - 2x^2$
(d) $16x^2 - 9y^2 = 144$   (e) $9x^2 + 4y^2 = 36$   **5**(a) $x = \dfrac{ct}{1 - mt}$, $y = \dfrac{c}{1 - mt}$

(b) $x = \dfrac{t}{\sqrt{t^2 - 1}}$, $y = \dfrac{1}{\sqrt{t^2 - 1}}$   (c) $x = \dfrac{t}{at^2 + b}$, $y = \dfrac{1}{at^2 + b}$

## 8.3

**1**(a) $\pi a^2$  (b) $\dfrac{a^2\pi}{8}$  (c) $\dfrac{7}{384}\pi^3 a^2$  (d) $\dfrac{a^2}{8}(9\pi + 16)$  (e) $\dfrac{\pi a^2}{6}$  **2** $\tfrac{1}{8}\pi a^2$

**3** $\tfrac{1}{2}\pi a^2$  **4** 1:7  **5**(a) $\tfrac{1}{4}\pi a^2$  (b) $\tfrac{1}{8}\pi a^2$  (c) $\dfrac{1}{12}\pi a^2$  (d) $\dfrac{1}{16}\pi a^2$  **6** $\dfrac{a^2}{2}$

## 8.4

### Exercise A

**2**(a) $-\dfrac{8}{9} < y < 0$  (b) $y > \dfrac{1}{8}$  (c) $y = 2$ or $y = 0$  **3**(a) $(1, 1)$ maximum, $(-1, -1)$ minimum  (b) $(-1, 0)$ minimum  (c) $\left(\dfrac{5}{3}, \dfrac{17}{4}\right)$ maximum; $(0, 0)$, $(\sqrt{3}, \tfrac{1}{2}\sqrt{3})$, $(-\sqrt{3}, -\tfrac{1}{2}\sqrt{3})$  **4**(a) symmetrical, $x = 0$  (b) symmetrical, any straight line through $(1, 0)$  (c) antisymmetrical, $y = -1$ and $x = 0$  (d) neither  (e) neither  (f) symmetrical $y = 0$

### Exercise B

**2**(a) $y < 0$  (b) $-3 < y < 0$  (c) $\dfrac{-2 - \sqrt{396}}{49} < y < \dfrac{-2 + \sqrt{396}}{49}$

**3**(a) $(0, 0)$ maximum, $(1, \tfrac{1}{2})$ minimum  (b) $(1, \tfrac{1}{3})$ maximum  (c) $(-1, 0)$ minimum  **4**(a) antisymmetrical $x = 0$, $y = 0$  (b) neither  (c) symmetrical $x = 0$  (d) antisymmetrical $y = 0$, $x = 0$  (e) neither  (f) symmetrical $x = 1$

## 8.5

**2** $(1, 1)$ minimum, $(\sqrt[3]{2}, 0)$ point of inflexion  **3** $\left(\dfrac{4}{5}, \dfrac{4^9}{5^5}\right)$ maximum, $(4, 0)$ minimum  **4** $(\sqrt{2}, 3 + 2\sqrt{2})$ minimum, $(-\sqrt{2}, 3 - 2\sqrt{2})$ maximum  **5** $(5^{-1/4}, 4 \times 5^{-1/4})$ minimum, $(-5^{-1/4}, -6 \times 5^{-\frac{1}{4}})$ maximum, $(0, 0)$ point of inflexion  **6** asymptotes $x = -1$, $y = 4$; zero $(-\tfrac{2}{3}, 0)$  **7** $x = 0.34$  **8** $x = 0.21$  **9**(a) $x = 1$, $y = 2$  (b) $x = 1.85$, $y = 5.55$  (c) $x = 1.47$, $y = 0.33$

## 9.2

**1**(a) $x > 6$  (b) $x > -\dfrac{7}{6}$  (c) $-\tfrac{1}{2} < x < 0$  (d) $x \geqslant \dfrac{2}{3}$  (e) $-1 \leqslant x \leqslant 0$, and $x \geqslant 1$  (f) $x > -2\tfrac{1}{2}$ or $x < -3$  (g) $x < 1$ or $x \geqslant \dfrac{7}{4}$  (h) $x < -5$ or $-1 \leqslant x \leqslant 1$  (i) $x \geqslant \tfrac{1}{2}$  **2**(a) $x < -1$ or $x > 1$  (b) $-1 < x < 5$  (c) $-\tfrac{1}{3} < x < 1$  (d) $\tfrac{1}{2}(1 - \sqrt{7}) \leqslant x \leqslant \tfrac{1}{2}(1 + \sqrt{7})$  (e) $\tfrac{1}{4}(-1 - \sqrt{5}) < x < 0$, $x > \tfrac{1}{4}(-1 + \sqrt{5})$  (f) no values

(g) $x \leqslant -1$, $-\frac{1}{7} < x \leqslant 2$    (h) all $x$    3(a) $x > 5$ or $x < -1$
(b) $x < -\frac{1}{2}$    (c) $x < \frac{2}{3}$    (d) $\frac{11}{5} < x < \frac{13}{3}$    (e) $x > -\frac{11}{8}$ or $x < -\frac{19}{4}$
(f) $x < -\frac{1}{2}$    (g) $1 < x < 2$    (h) $1 < x \leqslant 3$
(i) $x < -3\frac{1}{2}$, $x > -\frac{1}{2}$

## 9.3

### Exercise A
1(a) $x < -3$, $x > 3$    (b) $x > 2$, $x < -4$    (c) $x > \frac{1}{2}$, $x < -2$
(d) $\frac{5}{3} < x < \frac{7}{2}$    (e) $x < -\frac{7}{5}$, $x > -\frac{5}{7}$    2(a) $x < -2$, $x > 1$    (b) $x < 1$, $x > 1\frac{1}{2}$    (c) $-1 < x < 2$    (d) all $x$    (e) $x < -\frac{2}{3}$, $x > \frac{1}{2}$    3(a) $k > 1$
(b) $k > 2\frac{1}{8}$    (c) $k > 1$    (d) $k > 8\frac{3}{8}$    (e) $k > 0$, (f) $k > -1\frac{3}{4}$
4(a) min of 7    (b) max of $-\frac{7}{8}$    (c) min of $\frac{11}{12}$    (d) max of $-\frac{1}{5}$
(e) max of $-\frac{5}{8}$    5(a) $-3 < x < -2$, $x > 1$    (b) $x < -4$, $7 < x < 8$
(c) $x \leqslant -2$, $1 \leqslant x \leqslant 3$    (d) $-4 \leqslant x \leqslant 3$, $x \geqslant 5$
(e) $-1\frac{1}{2} < x < -1$, $0 < x < 2$    (f) $x < -7$, $-2\frac{1}{2} \leqslant x \leqslant 3$, $x > 6$

### Exercise B
1(a) $-4 < x < 4$    (b) $1\frac{1}{2} < x < 2$    (c) $x < -5$, $x > \frac{1}{2}$
2(a) $-1 < x < 1$    (b) $-3 < x < \frac{1}{2}$    (c) $x < -1\frac{1}{2}$, $x > 2$    3(a) none
(b) $k < -\frac{9}{16}$    (c) $k < -3\frac{3}{8}$    (d) $k > \frac{9}{8}$    (e) $-2\sqrt{2} < k < 2\sqrt{2}$
(f) $k < -\frac{1}{2}$    4(a) true for all $x$, min of 3    (b) not true for all $x$, max of $\frac{9}{8}$
(c) not all $x$, max of $\frac{25}{16}$    (d) true for all $x$, min of $\frac{7}{16}$    (e) not all $x$,
max of 9    5(a) $-6 < x < -3$, $x > 1$    (b) $x \leqslant -\frac{4}{3}$, $-1 \leqslant x \leqslant 1\frac{1}{2}$
(c) $-\frac{3}{4} \leqslant x \leqslant -\frac{1}{2}$, $x \geqslant \frac{2}{3}$    (d) $x \leqslant -1$, $-\frac{2}{3} \leqslant x \leqslant 3$
(e) $x < -1$, $-\frac{1}{2} < x < 0$, $x > 2$    (f) $x < -3$, $-1 < x < 1$
(g) $-2 \leqslant x \leqslant -\frac{1}{3}$, $\frac{1}{2} < x < \frac{2}{3}$

## 9.4
1(a) $0 < k < 2$    (b) $-8 < k < 0$    (c) $k > \frac{1}{3}$    (d) no values of $k$
2(a) $k \geqslant 1$, $x = -1$, $k = 1$    (b) $k \geqslant \frac{13}{4}$, $x = 1\frac{1}{2}$, $k = \frac{13}{4}$    (c) $k \leqslant -\frac{1}{2}$
and $k \geqslant \frac{1}{2}$, $x = -\frac{1}{4}$, $k = \pm\frac{1}{2}$    (d) $1 \leqslant k \leqslant 5$, $x = -\frac{1}{4}(k+1)$, $k = 1$ or 5
3(a) yes (if $x, y \geqslant 0$)    (b) yes    (c) no    4(a) no values
(b) $0 \leqslant x \leqslant 3$    (c) $1 \leqslant x \leqslant 2$    (d) $x < -1$ or $x > 3$
5(a) $-2-\sqrt{3} < x < -2$ and $x > \sqrt{3}-2$    (b) $\frac{1}{2}(-3-\sqrt{5}) < x \leqslant -2$
or $-1 \leqslant x < \frac{1}{2}(-3+\sqrt{5})$    (c) $\frac{27-\sqrt{129}}{10} < x < \frac{27+\sqrt{129}}{10}$

6(a) $\frac{1}{2} < x \leqslant 4$  (b) $-\frac{3}{2} < x < \frac{1}{2}(-3-\sqrt{19})$, $x > \frac{1}{2}(-3+\sqrt{19})$

## 10.2

### Exercise A

1(a) 5  (b) 17  (c) $\sqrt{82}$  (d) 7.872  (e) $2\sqrt{5}a$  (f) $\left|\sqrt{2}\left(1-\frac{1}{t}\right)\right|$
(g) $\sqrt{2(t^2+4)}$  2(a) $(2\frac{1}{2}, 4)$  (b) $(6, -8\frac{1}{2})$  (c) $(-4\frac{1}{2}, -7\frac{1}{2})$
(d) $(3.55, -1.7)$  (e) $(0, 0)$  (f) $\dfrac{t^2+1}{2t}, \dfrac{t^2+1}{2t}$  (g) $\{3t+2, \frac{5}{2}t-2\}$

3 $(3, \frac{43}{6})$  4 $(4a, 4a), (7a, 6a)$  5(a) $\dfrac{4}{3}$  (b) $\dfrac{15}{8}$  (c) 9  (d) $-0.479$

(e) 2  (f) $-1$  (g) $\dfrac{t+2}{t-2}$  6(a) $-\dfrac{3}{4}$  (b) $-\dfrac{8}{15}$  (c) $-\dfrac{1}{9}$  (d) 2.09

(e) $-\frac{1}{2}$  (f) 1  (g) $\dfrac{2-t}{2+t}$  7 $(2, 3)$  8 $\sqrt{13}$  9(a) $y = 6x+3$
(b) $y = 4x$  (c) $x+y = 1$  (d) $3y = x+12$  (e) $3y = 2x-8$
(f) $y = 3x+3$  (g) $2y = x-2$  (h) $16y = 10x-17$
(i) $\sqrt{3}(2y-7) = 2x+3$  (j) $342y = 940x \pm 1000$
(k) $259y - 966x = \pm 3400$  10(a) $\dfrac{9}{5}$  (b) $\dfrac{27}{\sqrt{65}}$

(c) $\dfrac{41t}{13} - 1$  (d) $\left|\dfrac{b^2-a^2}{\sqrt{a^2+b^2}}\right|$  11(a) $25°\,1'$  (b) $32°\,10'$  (c) $45°\,14'$

### Exercise B

1 $\left(-\dfrac{14}{4}, \dfrac{51}{4}\right), \left(-\dfrac{67}{7}, \dfrac{114}{7}\right)$  2 $(1, 7)$  3 $\dfrac{19}{8}, (1, 1\frac{1}{2}), \left(-\dfrac{3}{8}, -4\right)$
4(a) $y = 2x+3 = 0$  (b) $4y+3x = 9$  (c) $2y = 3x$  (d) $5y+7x = 39$
(e) $22y = 12x+126$  (f) $x+y = 5$  (g) $3x = 3 \pm \sqrt{120}$
(h) $y = 2x + 2\sqrt{5} - 7$ or $y = 2x - 2\sqrt{5} - 7$
(i) $1311x = 527y + 5242$  (j) $2y + 2\sqrt{3x} = 3$  (k) $x+7y = \pm 9\sqrt{50}$
5(a) $\dfrac{11}{\sqrt{17}}$  (b) $\dfrac{25}{\sqrt{13}}$  (c) 0  (d) $\dfrac{7a}{\sqrt{10}}$
6(a) $(3\sqrt{2}-5)x + (5-4\sqrt{2}y) = 2\sqrt{2}+2$
(b) $(\sqrt{5}-3)x + (\sqrt{5}-1)y = 12\sqrt{5}-6$
(c) $(\sqrt{17}+\sqrt{5})x + (2\sqrt{17}-4\sqrt{5}y) = 7\sqrt{17}-3\sqrt{5}$

## 10.3

### Exercise A

2(a) $(5, 14)$  (b) $(8, 4)$
(c) $(\frac{1}{2}(2+\sqrt{14}), \frac{1}{2}(2-\sqrt{14})), (\frac{1}{2}(2-\sqrt{14}), \frac{1}{2}(2+\sqrt{14}))$

(d) $(2(\sqrt{26}+2), \frac{1}{2}(\sqrt{26}-2)), (2(2-\sqrt{26}), -\frac{1}{2}(\sqrt{26}+2))$
(e) no points of intersection  3(a) $y = x+1$  (b) $(y-4)^2 = 9(x-1)$
(c) $x^2 + y^2 = 1$  (d) $y^2 = 4ax$  (e) $9x^2 - 16y^2 = 144$  (f) $y = (x^2+1)^2$
5(a) 3 or $-1$  (b) $\frac{4}{3}$ or 1  (c) 2 or $-\frac{1}{2}$  6(a) $y = 16x - 14$,
$16y + x = 290$  (b) $y = x - 1, y + x = 3$  (c) $2y + x = 2\sqrt{5}, y = 2x + \sqrt{5}$
(d) $24y + 11x + 26 = 0, 11y + 70 = 24x$  (e) $4y + x = 4, 2y - 8x + 15 = 0$
(f) $y + (\sqrt{2}+1)x = \frac{1}{4}\pi + 1 + \sqrt{2}, y - (\sqrt{2}-1)x = \frac{1}{4}\pi - 1 + \sqrt{2}$
(g) $ay\tan t - bx\sec t + ab = 0, by\sec t + ax\tan t + (a^2 - b^2)\sec t \tan t = 0$

7 $3y - 6 = (12 \pm \sqrt{78})(x-1)$  8 $\dfrac{1}{\sqrt[5]{6}}$  9 $(-12, 0), (0, 6), 36$

10(a) $h = \pm k$  (b) $8h + 15k - 1 = \pm 136$  (c) $k^2 = 4ah$
(d) $15k^2 + 15h^2 - 28ah - 62ak + 75a^2 = 0$  (e) $h - 2k + 4 = \pm 2\sqrt{5}$
(f) $|3k - 4h - 1| = 24$  (g) $3 - k = t(h+2)$  11(a) $18x + 12y + 69 = 0$
(b) $2y^2 + 2x^2 - 24y - 22x + 125 = 0$
(c) $32x^2 + 32y^2 - 36y - 108x + 8xy - 45 = 0$
(d) $2x^2 = 27y^3$  (e) $(y-1)(y+6) + (3+x)(2+x) = 0$
(f) $2y - \sqrt{3} = 2\sqrt{3}x$

**Exercise B**
1(a) 0 or $\frac{2}{3}\pi$  (b) $26° 34', 153° 26', 206° 34', 333° 26'$  (c) $\pm 3$
2(a) $10y = 3x + 26, 3y + 10x = 95$  (b) $5y + x = 6\sqrt{2}, y = 5x - 4\sqrt{2}$
(c) $2y = 9x - 27, 9y + 2x = 261$  (d) $y + 36x + 75 = 0$,
$108y + 1615 = 3x$  (e) $y + x = \sqrt{a^2+b^2}, y = x - \dfrac{b^2 - a^2}{\sqrt{a^2+b^2}}$
(f) $y + 4x + 8 = 0$ or $y + x = 4, 4y + 15 = x$ or $y = x$
3 $\left(\dfrac{134}{649}, \dfrac{261}{649}\right)$  4 $y = 4 \pm \sqrt{3}x$  5 $\sqrt{1 + (1-p)^2}, 1$
6(a) $h^2 + k^2 = 36$  (b) $2h + 3k - 7 = \pm\sqrt{13(h^2+k^2)}$
(c) $(3k+2)(k+6) + (3h+1)(h+3) = 0$  (d) $k + t_1 h = 2t_1^3 + 5t_1 - 3$
(e) $(12h - 6k - 115)^2 = 256\{(h-9)^2 + (k-2)^2\}$  (f) $h^2 + k^2 = 9$
7(a) $5x + 3y - 37 = 0$  (b) $8x^2 + 8y^2 + 94y - 60x + 329 = 0$
(c) $x^2 + y^2 - 12x - 10y + 36 = 0$  (d) $x^2 + y^2 = 9$
(e) $(x-1)^2 + (2y+1)^2 = 3$  (f) $9x + 12y = 5$  (g) $y^2 = a(x-a)$

## 10.4

**Exercise A**
1(a) $x^2 + y^2 = 16$  (b) $(x-6)^2 + (y-2)^2 = 25$
(c) $4(2x-1)^2 + (4y-1)^2 = 192$  (d) $(50x+10)^2 + (50y-70)^2 = 34596$
(e) $(x-1)^2 + (y-3)^2 = 5$  2(a) $a = -g, b = -f, r^2 = g^2 + f^2 - c$
3(a) yes, $(-1, -1), 1$  (b) no  (c) yes, $(-\frac{1}{4}, 0), \frac{1}{4}$  (d) no
(e) yes, $(-1, -2), \sqrt{\dfrac{23}{3}}$  (f) yes, $(-2, 0), \sqrt{6}$  (g) no  (h) no
4(a) $x^2 + y^2 = 169$  (b) $(x+2)^2 + (y-4)^2 = 50$
(c) $(x-5)^2 + (y-5)^2 = 10$  (d) $(x-3)^2 + (y-3)^2 = 9$

(e) $7x^2 + 7y^2 - 203x - y - 520 = 0$  (f) $(2y-1)^2 + (2x-1)^2 = 10$
(g) $(x-4)^2 + (y+1)^2 = 16$  (h) $4(x-6)^2 + (2y-13)^2 = 61$  6(a) $y = 0$
(b) $\sqrt{2}x + \sqrt{6}y = 8$  (c) $13y - 2x = 37$  (d) $x = 2$  7(a) 2
(b) $\sqrt{23}$  (c) $\sqrt{78}$  (d) $\sqrt{82}$  (e) $\sqrt{214}$
8  $8x^2 + 8y^2 + 1309x - 1253y + 1074 = 0$
9  $x^2 - 4x + y^2 - 2y + 1 - \frac{1}{7}(12 \pm \sqrt{60})(x^2 + y^2 - 9) = 0$
11  $40x^2 + 40y^2 + 64x + 88y + 7 = 0, \frac{1}{2}\sqrt{\frac{71}{10}}, \left(-\frac{4}{5}, -\frac{11}{10}\right)$

12(a) yes  (b) no  (c) yes  (d) no  (e) yes

**Exercise B**
1(a) $x^2 + y^2 = 289$  (b) $11x^2 + 11y^2 - 207x + 3y + 182 = 0$
(c) $(2x - (9 + \sqrt{17}))^2 + (2y - (9 + \sqrt{17}))^2 = 36$
or $(2x - (9 - \sqrt{17}))^2 + (2y + (9 - \sqrt{17}))^2 = 36$  (d) $2x^2 + 2(y+1)^2 = 9$
(e) $5(8x - 7)^2 + 320(y - 2)^2 = 36$  2(a) $y + 3x + 1 = 0$  (b) $5x + y + 11 = 0$
3(a) $4y + 3x = 15$ and $4y = 3x - 15$  (b) $y = 0, x = 4$
(c) $x \pm 2\sqrt{2}y = 8 \pm 2\sqrt{2}$  4 $(-10, 5)$  5 $c < -\sqrt{45}$ or $c > \sqrt{45}, \sqrt{45}$
6(b) $4y = x \pm 3\sqrt{17}$  (c) $5y = x \pm \sqrt{156}$  7(a) $\sqrt{59}$  (b) $\sqrt{\frac{122}{3}}$
(c) $\sqrt{70}$  8(a) $127x^2 + 127y^2 - 800x - 932y - 272 = 0$
(b) $73x^2 + 73y^2 + 100x + 182y - 203 = 0$  (c) $4x^2 + 4y^2 + 4x + 8y - 11 = 0$
(d) $3x^2 + 3y^2 + 2y - 8 = 0$  9 $12x + 16y = 1$  10 $-\frac{1}{8}, -\frac{13}{2},$

$(\frac{1}{8}, -1), \frac{\sqrt{481}}{8}$  12(a) yes  (b) yes  (c) no  (d) no

## 10.5

**Exercise A**
1(a) yes, $(0, 0), (1, 0), x = -1$  (b) yes, $(0, 0), (2, 0), x = -2$  (c) no
(d) yes, $(0, 0), (0, 2), y = -2$  (e) yes, $(0, 2), (\frac{1}{4}, 2), x = -\frac{1}{4}$
(f) yes, $(-\frac{1}{4}, -1), (\frac{3}{4}, -1), x = -1\frac{1}{4}$  (g) no  (h) yes,
$\left(\frac{79}{64}, -\frac{1}{8}\right), \left(\frac{95}{64}, -\frac{1}{8}\right), x = \frac{63}{64}$  (i) yes, $(0, 0), \left(\frac{b}{4a}, 0\right), x = -\frac{b}{4a}$
2(a) $y^2 = 8x$  (b) $y^2 = 16(x - 1)$  (c) $y^2 = 4(x - 2)$
(d) $(y-1)^2 = 8(x+1)$  (e) $(y-3)^2 = 4(x-2)$  (f) $(x-1)^2 = 16(y-3)$
4  $py = x + ap^2, y + px = 2ap + ap^3$  5 $\pm 1$  10 $y = \frac{2}{m}$  11 $(\frac{9}{2}a, 0)$
12  $a^2(1+q^2)(p-q)$  13 $2y = x + 4, y = x + 1$  14 $12y + 8x + 45 = 0$
15  $3y = 4$

**Exercise B**
1(a) yes, $(0, 0), x = -\frac{1}{2}, (\frac{1}{2}, 0)$  (b) yes, $(0, 0), x = -2, (2, 0)$  (c) yes, $(0, 0), x = \frac{1}{4}, (\frac{1}{4}, 0)$  (d) yes, $(0, 0), y = -1, (0, 1)$  (e) yes, $(1, 0), y = 1,$

132

(0, −1)  (f) no  (g) yes, (−17, −4), $x = -17\frac{1}{4}$, $(-16\frac{3}{4}, -4)$  (h) yes, $(2a, -a)$, $y = -2a$, $(2a, 0)$  2(a) $y^2 = 16x$  (b) $y^2 = 8(x-1)$
(c) $(y-2)^2 = 4(x-1)$  (d) $(y-2)^2 = 8(x+1)$  (e) $(x+2)^2 = 12y$
3  $4a$  4  $2y = x+4$, $2x+y = 12$, $(9, -6)$  5  $y = 4$
7  $x^2 + y^2 - 16x - 4y + 2xy + 10 = 0$
8  $\sqrt{6}y = 3x+6$, $3x + \sqrt{6}y + 6 = 0$  9  $(0,0)$, $90°$, $(4,4)$, $\tan^{-1}\frac{3}{4}$
10(a) $2a^2t(t^2+1)$  (b) $2a^2t(t^2+1)$

## 10.6

1  $(-3, 0)$, $x = -12$  4(a) $\frac{\sqrt{7}}{4}$, $(\sqrt{7}, 0)$ and $(-\sqrt{7}, 0)$, $x = \pm\frac{16}{\sqrt{7}}$

(b) $\frac{\sqrt{5}}{3}$, $\left(0, \frac{2\sqrt{5}}{3}\right)$ and $\left(0, -\frac{2\sqrt{5}}{3}\right)$, $y = \pm\frac{6}{\sqrt{5}}$  5(a) $\frac{x^2}{16} + \frac{y^2}{15} = 1$

(b) $4\frac{x^2}{9} + y^2 = 1$  (c) $\frac{x^2}{16} + \frac{y^2}{12} = 1$  7(a) $2y - 3x = 12$, $3y + 2x = 5$
(b) $x + 2\sqrt{3}y = 4$, $4\sqrt{3}x - 2y = 3\sqrt{3}$  (c) $5x - 21y = 34\sqrt{2}$, $5\sqrt{2}y + 21\sqrt{2}x = 6$  (d) $bx + 3ay = 11ab$, $by = 3ax + 3b^2 - 6a^2$
8  $\pm\sqrt{24}$  9  $y = 3x \pm \sqrt{59}$, $3y + x = \pm\frac{3}{\sqrt{59}}$  13  $(-4, 0)$, $x = -1$

16(a) $\frac{5}{3}$, $(\pm 5, 0)$, $x = \pm\frac{9}{5}$  (b) $\sqrt{\frac{5}{2}}$, $(\pm\sqrt{5}, 0)$, $x = \pm\frac{2\sqrt{5}}{5}$

(c) $\frac{\sqrt{65}}{4}$, $(\pm\sqrt{65}, 0)$, $x = \pm\frac{16}{\sqrt{65}}$  17(a) $x = \pm\sqrt{5}y$

(b) $\sqrt{5}x = \pm 2y$  (c) $3x = \pm 2y$  18(a) $8x^2 - y^2 = 32$, $x = 2\sec\theta$, $y = 4\sqrt{2}\tan\theta$  (b) $3x^2 - y^2 = 12$, $x = 2\sec\theta$, $y = \sqrt{12}\tan\theta$
(c) $3x^2 - y^2 = 27$, $x = 3\sec\theta$, $y = 3\sqrt{3}\tan\theta$  (d) $45x^2 - 5y^2 = 18$, $x = \frac{\sqrt{10}}{5}\sec\theta$, $y = \frac{3\sqrt{10}}{5}\tan\theta$  (e) $64x^2 - y^2 = 576$, $x = 3\sec\theta$, $y = 24\tan\theta$  19(a) $x + y = \frac{3\sqrt{3}}{2}$, $y - x + \frac{5\sqrt{3}}{2} = 0$
(b) $5x - \sqrt{3}y + 5 = 0$, $5y + \sqrt{3}x + 29\sqrt{3} = 0$  (c) $x = 5$, $y = 0$
(d) $9\sqrt{2}x - 4\sqrt{3}y = 6$, $9\sqrt{2}y + 4\sqrt{3}x = 13\sqrt{6}$  20  $\pm\frac{3}{2}\sqrt{15}$
21(a) $c^2 < 7$  (b) $c^2 > 7$  (c) $c^2 = 7$; $y = x \pm \sqrt{7}$, $\left(\pm\frac{16}{\sqrt{7}}, \pm\frac{9}{\sqrt{7}}\right)$,

$y + x = \pm\frac{25}{\sqrt{7}}$  22  $5y + 26x = 36$, $y = 2x$  23(a) $8x + y = 16$, $8y = x + 63$
(b) $4x + y = 4\sqrt{3}$, $8y = 2x + 15\sqrt{3}$

133

(c) $x + 25y = 30$, $5y = 125x - 1872$
(d) $yt_1 + \dfrac{x}{t_1} = 2c$, $t_1 y = t_1^3 x + c - ct_1^4$  (e) $xy_1 + x_1 y = 2c^2$,

$y - y_1 = \dfrac{x_1}{y_1}(x - x_1)$  24  $-\dfrac{1}{27}, \left(\dfrac{80}{9}, -80\right)$  26  $\pm 2b\sqrt{-m}$

27  $x = 2$   28  $y = -2x$

29  $(x+y)^2 + 2xy = \dfrac{2\sqrt{2}}{3}(x+y)xy$

## 10.7

1  $9\frac{1}{2}$   2  $(3, \frac{8}{3})$   3(a) $\left(-\dfrac{3}{13}, \dfrac{15}{13}\right)$   (b) $\left(\dfrac{159}{65}, \dfrac{2}{65}\right)$

4  $\dfrac{2x+y-4}{\sqrt{5}} = \pm \dfrac{5x+2y-6}{\sqrt{29}}$   7  $y = \pm\sqrt{3}x - 4$   8  $\pm\dfrac{3}{\sqrt{2}}$

10  $y^2 = 2a(x - \frac{1}{4}a)$   14  $\left(\dfrac{a\cos\frac{1}{2}(\theta_1+\theta_2)}{\cos\frac{1}{2}(\theta_1-\theta_2)}, \dfrac{b\sin\frac{1}{2}(\theta_1+\theta_2)}{\cos\frac{1}{2}(\theta_1-\theta_2)}\right)$

18(a) $(x+2)^2 + (y+3)^2 = a^2$   (b) $(y+2)^2 = 4a(x-1)$

(c) $\dfrac{x^2}{a} + \dfrac{(y-3)^2}{b} = 1$   (d) $y = 2x + 1$   (e) $(x-5)(y-2) = 9$

(f) $(x+2)(x+4) = y$   (g) $(y-2)^2 = x+3$

(h) $y + 4 = \left(\dfrac{x+5}{x+3}\right)^2$   19(a) $(a, b)$ circle   (b) $(1\frac{1}{2}, 1\frac{1}{4})$ circle   (c) $(-2, 0)$ ellipse   (d) $(-1, 1)$ rectangular hyperbola   (e) $(-3, -2)$ hyperbola   (f) $(0, 0)$, pair of straight lines

## 1

**Exercise A**
1(a) $2\frac{1}{2}$, $-1$   (b) $-1.5$, $-2.3$   2  $2, -1$   3  $1, 1$   4  $2.5, 1.5$   5(a) $3.1, -1.3$   (b) $0.2, 0.25$   (c) $7.5, 3$   (d) $0.2$   (e) $-1, 1$   (f) $3.5$

**Exercise B**
1  $5, 2$   2  $3, 5$   3  $1.4, 3$   4(a) $4.6, 5.8$   (b) $4.9, 1.75$   (c) $0.25, 0.3$
(d) $2.1, 2.1$   (e) $3, 2$

## 12.2

1(a) $7+i$, $1-2i$, $1$   (b) $6$, $-\sqrt{3}-2i$, $-2+3\sqrt{2}i$   (c) $6+17i$, $10-6i$, $-2-6i$   (d) $4$, $34-13i$, $6-17i$   (e) $-7+24i$, $-3-4i$, $8+6i$
(f) $\dfrac{-7+24i}{25}$, $\dfrac{1-5i}{2}$, $3+2i$   (g) $5i$, $\dfrac{9-13i}{5}$, $-2+2i$   2(a) $1+i$   (b) $5$
(c) $3+3\sqrt{3}i$   (d) $-4$   (e) $-1+\sqrt{3}i$   (f) $2-2i$   (g) $3$

(h) $-5\sqrt{2}+5\sqrt{2}i$  (i) $-7$  (j) $-4-4\sqrt{3}i$  3(a) $(1, \frac{1}{6}\pi)$
(b) $(1, \frac{1}{4}\pi)$  (c) $(1, \frac{2}{3}\pi)$  (d) $(1, -\frac{5}{6}\pi)$  (e) $(\sqrt{2}, -\frac{3}{4}\pi)$  (f) $(2, -\frac{1}{6}\pi)$
(g) $(5, \tan^{-1}\frac{4}{3})$  (h) $(6, -\frac{1}{2}\pi)$  (i) $(13, \tan^{-1}\frac{5}{12})$  4(a) $(6, \frac{1}{2}\pi)$, $6i$
(b) $(8, \pi)$, $-8$  (c) $(24, \frac{5}{6}\pi)$, $-12\sqrt{3}+12i$  (d) $(4, \frac{3}{4}\pi)$, $-2\sqrt{2}+2\sqrt{2}i$
(e) $(25, -\frac{2}{3}\pi)$, $-25-\dfrac{25\sqrt{3}}{2}i$

5(a) $(3, \frac{1}{4}\pi)$, $\dfrac{3}{\sqrt{2}}+\dfrac{3}{\sqrt{2}}i$  (b) $(2, \frac{1}{2}\pi)$, $2i$  (c) $(2\frac{1}{2}, \pi)$,
$-2\frac{1}{2}$  (d) $(4, \frac{1}{6}\pi)$, $2\sqrt{3}+2i$  (e) $(4, \frac{2}{3}\pi)$, $-2+2\sqrt{3}i$

## 12.3

**Exercise A**

1(a) $1\pm 2i$  (b) $-1\pm\sqrt{2}i$  (c) $\frac{1}{2}\pm\dfrac{\sqrt{7}}{2}i$  (d) $-\frac{1}{4}\pm\dfrac{\sqrt{31}}{4}i$  2 $1-2i$,
$3+2i$, $-3i$, $-4+5i$, $\cos\theta-i\sin\theta$  3 $(z-1)(z^2+z+1)=0$, $1$,
$-\frac{1}{2}\pm\dfrac{\sqrt{3}}{2}i$, each is the square of the other  4(a) $1$, $-\frac{1}{2}\pm\dfrac{\sqrt{7}}{2}i$
(b) $2$, $-\frac{1}{2}\pm\frac{3}{2}i$  (c) $3$, $-\frac{1}{2}\pm\dfrac{\sqrt{11}}{2}i$  (d) $\frac{1}{2}$, $\frac{1}{2}\pm\dfrac{\sqrt{3}}{2}i$  5 $2$, $-2$, or $\pm 2i$

**Exercise B**

1(a) $2\pm i$  (b) $-\frac{1}{2}\pm\dfrac{\sqrt{11}}{2}i$  (c) $\dfrac{3\pm\sqrt{3}i}{2}$  (d) $\dfrac{3\pm\sqrt{3}i}{6}$  2 $2-i$, $2+3i$,
$-2-5i$, $4$, $\cos\phi+i\sin\phi$  3 $(z+1)(z^2-z+1)=0$, $-1$, $\frac{1}{2}\pm\dfrac{\sqrt{3}}{2}i$, the three
cube roots of unity  4(a) $-1$, $1\pm i$  (b) $-2$, $1\pm i$  (c) $4$, $-2\pm\sqrt{2}i$
(d) $\frac{1}{2}$, $-1\pm i$  5 $(z-1)(z+1)(z^4+z^2+1)=0$, $z=\pm 1$, $z^2=\dfrac{-1\pm\sqrt{3}i}{2}$

## 12.4

1(a) square  (b) isosceles triangle  (c) parallelogram  (d) rectangle
6 rectangle  8 $5, 0, -2i$  10 $5+3i$

## 12.5

1(a) $(0, 0)$  (b) $(1, -2)$  (c) $(-1, 3)$  (d) $(1, -4)$  (e) $(2, -3)$
2(a) $(1, -2)$  (b) $(-1, 3)$  (c) $(10, -\frac{2}{3})$  3(a) $2x$
(b) $(x^2-y^2)+2ixy$  (c) $\dfrac{2x}{x^2+y^2}$  (d) $x^2+y^2$  (e) $\dfrac{x^2-y^2+2ixy}{(x^2+y^2)^2}$
4 $\pm(3+2i)$  5(a) $\pm(1+2i)$  (b) $\pm(2-5i)$  (c) $\pm(4-i)$

(d) $\pm(1-\sqrt{3}i)$  6(a) $(3, -2)$  (b) $(1, 4)$  (c) $(13, 1)$
(d) $(7, 1)$  (e) $(7, 5)$  (f) $(4, 3)$

## 12.6

2 $\cos 4\theta = 8\cos^4\theta - 8\cos^2\theta + 1$, $\sin 4\theta = 4\cos^3\theta \sin\theta - 4\cos\theta \sin^3\theta$
4(a) $4-i$  (b) $-4+i$  (c) $-4-i$  (d) $1+4i$  (e) $-1-4i$
5 $(\sqrt{2}, \pm\frac{1}{4}\pi)$, $5-\frac{3}{2}i$  6 $(2, \frac{3}{2}\pi)$  7 $\frac{1}{2}(1+i), \frac{1}{2}(3+i)$,
$2z^2 - 3(1+i)z + 3(1+i) = 0$  8 $\omega, \omega^2, -1, 1$  9 $a = 1, b = 5$;
$1\frac{1}{4} - \frac{1}{4}i$  10 $\dfrac{3-4i}{25}$, $\pm(2+i)$, trapezium  11 $3a+2b+c+2=0$,
$4a+b+c+11=0$, $a = -6, b = 3$.  12 $\pm 1 \pm i$

## 13.2

1(a) $5, 5$  (b) $4x + 2\delta x, 4x$  (c) $6x + 3\delta x + 1, 6x + 1$
(d) $3x^2 + 3x\,\delta x + (\delta x)^2, 3x^2$  (e) $2x + \delta x - 2, 2x - 2$
(f) $3x^2 + 3x\,\delta x + (\delta x)^2 + 2x + \delta x, 3x^2 + 2x$  2(a) $4x^3$  (b) $6x^5$
(c) $21x^2$  (d) $-6$  (e) $x^3$  (f) $6x - 6$  (g) $12x^2 + 1$
(h) $3x^2 - 4x + 3$  3(a) $-2x^{-3}$  (b) $-4x^{-5}$  (c) $-\dfrac{1}{x^2}$  (d) $-\dfrac{3}{x^4}$
(e) $-\dfrac{8}{x^3}$  (f) $-\dfrac{8}{x^5}$  (g) $-\dfrac{4}{x^3}+\dfrac{1}{x^2}$  (h) $-\dfrac{3}{x^4}-\dfrac{1}{x^2}+2x$
4(a) $\frac{1}{2}x^{-1/2}$  (b) $\dfrac{1}{3\sqrt[3]{x^2}}$  (c) $\frac{2}{3}x^{-1/3}$  (d) $\frac{3}{2}x^{1/2}$  (e) $\dfrac{2}{3}\dfrac{1}{\sqrt[3]{x}}$
(f) $-\dfrac{1}{3x\sqrt[3]{x}}$  (g) $-\dfrac{1}{4x\sqrt[4]{x}}$  (h) $-\dfrac{1}{2x\sqrt{x}} - \dfrac{1}{3x\sqrt[3]{x}}$  5(a) $3x^2 + 2$
(b) $4x^3 + 8x$  (c) $4x^3 + 3x^2 + 2x$  (d) $1 - \dfrac{2}{x^3}$  (e) $1 - \dfrac{3}{x^2}$
(f) $4x^3 - \dfrac{2}{x^3}$  (g) $2x - \dfrac{2}{x^3}$  (h) $3x^2 + 4x - 3$
(i) $\frac{5}{2}x^{3/2} + 2x + \frac{1}{2}x^{-1/2}$

## 13.3

1(a) $6(2x+3)^2$  (b) $20x(x^2+1)^4$  (c) $-\dfrac{6}{(3x+1)^3}$
(d) $3x^2\left\{1 - \dfrac{1}{(x^3-1)^2}\right\}$  (e) $\left\{4(x^2-4)+\dfrac{2}{(x^2-4)^3}\right\}2x$
2(a) $8x(x^2+2)$  (b) $24(2x+5)^3$  (c) $-\dfrac{12x}{(2x^2-3)^4}$
(d) $4x(x^2+6) - \dfrac{4x}{(x^2+6)^3}$  (e) $(3x^2+1)\{3(x^3+x)^2-1\}$
3(a) $12x^2(x^3+5)^3$  (b) $5(x^3+x^2+1)^4(3x^2+2x)$  (c) $-\dfrac{4x^3}{(x^4+4)^2}$

(d) $\frac{4}{3}x(2x^2-1)^{-2/3}$  (e) $-\frac{2x}{(2x-1)^{3/2}}$  (f) $\frac{1}{2}x(x^2-16)^{-3/4}$

4(a) $6x(x^2+2)^2$  (b) $4(x^2+2x+3)^3(2x+2)$  (c) $-\frac{3x^2}{(x^3-8)^2}$

(d) $\frac{x}{\sqrt{x^2+4}}$  (e) $-\frac{x}{(x^2+4)^{3/2}}$  (f) $\frac{x^2}{(x^3+8)^{2/3}}$  (g) $-\frac{12x}{(2x^2+3)^4}$

(h) $-\frac{3x}{(x^2+6)^{5/2}}$

## 13.4

1(a) $2\cos 2x$  (b) $-3\sin 3x$  (c) $\frac{1}{2}\sec^2\frac{1}{2}x$  (d) $2\cos(2x+1)$
(e) $-3\sin(3x+2)$  (f) $\frac{1}{2}\sec^2(\frac{1}{2}x+\frac{1}{4}\pi)$  (g) $2\cos(2x+\frac{1}{6}\pi)$
(h) $3\sec^2(3x+\frac{1}{2}\pi)$  (i) $-n\sin(nx+\pi)$  (j) $\sec^2(\frac{1}{2}x+2)$
(k) $-2\sin(\frac{1}{3}x+\frac{1}{4}\pi)$  (l) $\cos(\frac{1}{4}x-\frac{1}{2}\pi)$  2(a) $2\sin x\cos x$
(b) $-3\cos^2 x\sin x$  (c) $\frac{\sec^2 x}{2\sqrt{\tan x}}$  (d) $-\csc x\cot x$
(e) $-\csc^2 x$  (f) $2\tan x\sec^2 x$  (g) $-4\cos^3 x\sin x$
(h) $-2\csc^2 x\cot x$  (i) $-2\cot x\csc^2 x$
(j) $4\sin(2x+1)\cos(2x+1)$  (k) $-4x\cos(x^2+1)\sin(x^2+1)$
(l) $9x^2\tan^2(x^3+1)\sec^2(x^3+1)$  (m) $\frac{\cos(2x+\frac{1}{6}\pi)}{\sqrt{\sin(2x+\frac{1}{6}\pi)}}$  (n) $\frac{\sin 2x}{(\cos 2x)^{3/2}}$
(o) $-\frac{8\sec^2 4x}{\tan^3 4x}$  3(a) $3(1+\sin x)^2\cos x$  (b) 0  (c) $2\cos 2x$
(d) $-2\sin 2x$  (e) $\sec^2 x - \csc^2 x$  (f) $\sec x\tan x$  4(a) $2e^{2x}$
(b) $2e^{2x+1}$  (c) $2xe^{x^2}$  (d) $-2e^{-2x}$  (e) $-\frac{e^{1/x}}{x^2}$  (f) $2e^{1/2x}$  (g) $2xe^{x^2+1}$
(h) $-2xe^{-x^2}$  (i) $e^{\sin x}\cos x$  (j) $e^{\tan x}\sec^2 x$  (k) $2e^{\sin 2x}\cos 2x$
(l) $\frac{1}{2}e^{\tan\frac{1}{2}x}\sin^2\frac{1}{2}x$  5(a) $\frac{1}{x}$  (b) $\frac{1}{x+1}$  (c) $\frac{2}{x}$  (d) $\frac{2x}{x^2+1}$  (e) $\frac{2}{x}$
(f) $\frac{2x+2}{x^2+2x+3}$  (g) $\cot x$  (h) $2\cot 2x$  (i) $\frac{\sec^2 x}{\tan x}$
(j) $-2x\tan(x^2+1)$  (k) 1  (l) $\cos x$  6(a) $\frac{1}{\sqrt{1-x^2}}$  (b) $\frac{1}{1+x^2}$
(c) $-\frac{1}{\sqrt{4-x^2}}$  (d) $\frac{2}{4+x^2}$  (e) $\frac{2}{\sqrt{x-x^2}}$  (f) $\frac{4}{x^2+16}$
(g) $\frac{2}{\sqrt{9-4x^2}}$

## 13.5

**Exercise A**

1(a) $x^2\cos x+2x\sin x$  (b) $e^x(x+1)$  (c) $1+\log_e x$

137

(d) $e^x(\cos x - \sin x)$  (e) $\cos^2 x - \sin^2 x$
(f) $x^2 \cos x + 2x(1 + \sin x)$  (g) $2x \cos 2x + \sin 2x$
(h) $\dfrac{1 + 2x^2}{\sqrt{1 + x^2}}$  2(a) $\dfrac{2x \sin x - x^2 \cos x}{\sin^2 x}$  (b) $e^x\left(\dfrac{1}{x} - \dfrac{1}{x^2}\right)$
(c) $\dfrac{1 - \log x}{x^2}$  (d) $e^{-x}(\cos x - \sin x)$  (e) $\dfrac{e^x(\sin x - \cos x)}{\sin^2 x}$
(f) $\dfrac{(1 + \sin x)2x - x^2 \cos x}{(1 + \sin x)^2}$  (g) $\dfrac{\sin 2x - 2x \cos 2x}{\sin^2 2x}$  (h) $\dfrac{2}{(1 + x)^{3/2}}$
3(a) $e^x(\tan x + \sec^2 x)$  (b) $\dfrac{e^x(\tan x - \sec^2 x)}{\tan^2 x}$  (c) $e^x\left(\dfrac{1}{x} + \log x\right)$
(d) $\dfrac{e^x\left(\log x - \dfrac{1}{x}\right)}{(\log x)^2}$  (e) $\dfrac{1}{x}\sin x + \cos x \log x$  (f) $\dfrac{\sin x - x \cos x \log x}{x \sin^2 x}$

(g) $\dfrac{x}{\sqrt{1 - x^2}} + \sin^{-1} x$  (h) $\left\{\dfrac{x}{\sqrt{1 - x^2}} - \sin^{-1} x\right\}\dfrac{1}{x^2}$

4(a) $(1 + x^2)^2 \{(1 + x^2)\cos x + (1 - x^2)\sin x\}$
(b) $\dfrac{e^x(2x \sin x + x^2 - x^2 \cos x)}{\sin^2 x}$
(c) $\dfrac{\cos x (1 + \log x) + x \log x \sin x}{\cos 2x}$
(d) $e^{-x}((1 + x^2)\cos x - (1 - x^2)\sin x)$
(e) $\dfrac{e^x\{(x^2 - 5)\sec^2 x + (x^2 - 2x - 5)\tan x\}}{(x^2 - 5)^2}$
(f) $e^{-x} \sec x (\sec^2 x + \tan^2 x - \tan x)$

### Exercise B

1(a) $3x^2 \cos x - x^3 \sin x$  (b) $e^{-x}(1 - x)$  (c) $2x \log x + x$
(d) $e^{-x}(\cos x - \sin x)$  (e) $\sin x \sec^2 x + \sin x$
(f) $x^3 \sin x + 3x^2(1 - \cos x)$  (g) $2x \cos 2x - 2x^2 \sin 2x$
(h) $\dfrac{x^2}{2\sqrt{1 + x}} + 2x\sqrt{1 + x} = \dfrac{5x^2 + 4x}{2\sqrt{1 + x}}$  2(a) $\sec x(1 - x \tan x)$
(b) $e^{-x}(1 - x)$  (c) $\dfrac{x(2 \log x - 1)}{(\log x)^2}$  (d) $-e^{-x}(\sin x + \cos x)$
(e) $e^x(\cot x - \csc^2 x)$  (f) $\dfrac{(1 - \cos x)3x^2 - x^3 \sin x}{(1 - \cos x)^2}$
(g) $\dfrac{2x \cos 2x + 2x^2 \sin 2x}{\cos^2 2x}$  (h) $\dfrac{2x - x^3}{(1 - x^2)^{3/2}}$  3(a) $\dfrac{1}{x}\tan x + \sec^2 x \log x$
(b) $\dfrac{1}{x}\cot x - \csc^2 x \log x$  (c) $e^{-x}(\sec^2 x - \tan x)$
(d) $e^x(\sec^2 x + \tan x)$  (e) $\dfrac{e^x}{\sqrt{1 + x^2}}(1 + x + x^2)$  (f) $\dfrac{e^x}{(1 + x^2)^{3/2}}(1 - x + x^2)$

(g) $2x \tan^{-1} x + \dfrac{x^2}{1+x^2}$  (h) $\dfrac{1}{x^3}\left\{\dfrac{x}{1+x^2} - 2\tan^{-1} x\right\}$

4(a) $\dfrac{x^2 \sec^2 x}{1+x} + \dfrac{(x^2+2x)\tan x}{(1+x)^2}$  (b) $e^{-x}\sec^2 x\{(1-x)\cos x + x \sin x\}$

(c) $\operatorname{cosec} x(x + 2x \log x) - x^2 \cos x \operatorname{cosec}^2 x \log x$

(d) $-\dfrac{1}{x^4}(3x \cos x \sin 3x + x \sin x \cos 3x + 3 \cos x \cos 3x)$

(e) $\operatorname{cosec}^3 3x(\cos x \cos 2x \sin 3x - 2\sin x \sin 2x \sin 3x - 3 \sin x \cos 2x \cos 3x)$

(f) $\dfrac{x - (x^2+3)\tan^{-1} x}{x^4}$

## 13.6

1(a) $\dfrac{2a}{y}$  (b) $-\dfrac{9x}{16y}$  (c) $-\dfrac{y}{x}$  (d) $-\dfrac{x+1}{y+2}$  (e) $-\dfrac{2x+y}{x+2y}$

(f) $\dfrac{4}{3y^2}$  (g) $\dfrac{2-xy^2}{x^2 y}$  (h) $-\sqrt{\dfrac{y}{x}}$  (i) $\dfrac{y-x^2}{y^2-x}$

2(a) $-\dfrac{b}{a}\cot\theta$  (b) $-\dfrac{1}{t^2}$  (c) $-\dfrac{1+\cos\theta}{1+\sin\theta}$  (d) $\dfrac{t^2+1}{t^2-1}$  (e) $\dfrac{4}{3t}$

(f) $\dfrac{4(t-3)^2}{9(t-2)^2}$  (g) $-\tan\theta\, e^{\cos\theta-\sin\theta}$  (h) $-\tan^2\theta$  3(a) $\sqrt{\dfrac{1-x}{1+x}}\; \dfrac{1}{x^2-1}$

(b) $\dfrac{(x^2+1)^3}{\sqrt{2x+1}}\left\{\dfrac{6x}{x^2+1} - \dfrac{1}{2x-1}\right\}$  (c) $x^{\sin x}\left\{\dfrac{1}{x}\sin x + \cos x \log x\right\}$

(d) $x^{\tan x}\left\{\dfrac{1}{x}\tan x + \sec^2 x \log x\right\}$  (e) $\dfrac{x^4 \sin^4 x}{\cos^5 x}\left\{\dfrac{4}{x} + 4\cot x + 5\tan x\right\}$

(f) $x^x(1 + \log x)$  (g) $x^{e^x} e^x\left(\dfrac{1}{x} + \log x\right)$

(h) $\dfrac{(x+1)^5 \cos^3 x}{\sin^6 x}\left\{\dfrac{5}{x+1} - 3\tan x - 6\cot x\right\}$  (i) $100^x \log_e 100$

## 13.7

1(a) $\dfrac{8}{(1-3x)^2}$  (b) $8\sin 4x \cos 4x$  (c) $2\cos 2x\, e^{\sin 2x}$  (d) $\dfrac{2}{x^2-1}$

(e) $2(x^2+1)\sec^2 2x + 2x \tan 2x$  (f) $\dfrac{9x}{\sqrt{1+9x^2}}$

(g) $2\sin x \cos x (\cos^2 x - \sin^2 x) = \tfrac{1}{2}\sin 4x$  (h) $\pi \cos(\pi x) e^{\sin \pi x}$

(i) $2x \cos(x^2 + 1)$  (j) $\dfrac{(x^3 - 3x)}{(x^2-1)^{3/2}}$  (k) $-\tfrac{1}{2}\sec^2(x - \tfrac{1}{2}\pi)$  (l) $\dfrac{1}{x^2 - 1}$

2(a) $1 + \log x$  (b) $-4\operatorname{cosec}^2 2x \cot x$  (c) $2\log_e(10) 10^{2x}$  (d) $-\dfrac{2x^2+3}{x^4\sqrt{1+x^2}}$

(e) $\dfrac{1}{\sqrt{x^2+1}}$  (f) $\pi e^{\pi x}(\cos \pi x - \sin \pi x)$  (g) $\dfrac{3x^2}{x^3+1} - \dfrac{2x}{x^2+1}$

(h) $-4 \cot 2x \operatorname{cosec}^2 2x$  (i) $-5(1-2x)^{3/2}$  (j) $\dfrac{2e^{2x}}{e^{2x}-1}$

(k) $2x \sin x \cos 2x + x \cos x \sin 2x + \sin x \sin 2x$  (l) $\dfrac{1}{x} \log_{10} e$

(m) $-\dfrac{2}{x^3} + \dfrac{6}{x^4} - \dfrac{4}{x^5}$  (n) $\dfrac{1}{\sqrt{3-x^2+2x}}$  (o) $3(e^x \cos x)^3 (1 - \tan x)$

(p) $\dfrac{1}{\sqrt{x^2-2}}$  (q) $\operatorname{cosec} x$  (r) $-\dfrac{1}{\sqrt{a^2-x^2}}$  (s) $4 \cot 2x$

3  $y = \dfrac{10x^2 - 5x - 6}{(x^2-1)(2x-1)}$  4  $\dfrac{e^{-x}(x^4 + 4x - 1)}{(1-x^2)^2}, \dfrac{7}{9}e^2$  5  $\dfrac{2x+1}{2y-3}, -\dfrac{1}{7}, \dfrac{1}{7}$

6(a) $-\dfrac{b^2}{a^2}\dfrac{x}{y}, -\dfrac{b^2(b^2x^2 + a^2y^2)}{a^4 y^3}$  (b) $-\dfrac{x+1}{y+1}, -\dfrac{(y+1)^2 + (x+1)^2}{(y+1)^3}$

(c) $\dfrac{2y-x}{y-2x}, \dfrac{3x^2 + 3y^2 - 12xy}{(y-2x)^3}$  7 $y + ex = 0$  9 $\dfrac{1}{x-1} + \dfrac{2}{2x-3}$;

$-5, 18$  10(a) $-\dfrac{1}{x^2}$  (b) $\dfrac{1}{2\sqrt{x}}$  (c) $-\dfrac{1}{2x\sqrt{x}}$

11 $6y - 2\sqrt{3} = 2\sqrt{3}x$  12 $y = x - \tfrac{1}{2}\pi + 2$
13 $x^{2/3} + y^{2/3} = a^{2/3}, y + x \tan \alpha = a \sin \alpha$  14 $-\cot \tfrac{3}{2}\theta$

## 14.2

1(a) 3, 3  (b) 3, 7  (c) 4, 0  (d) 2, −2  (e) 0, $-\tfrac{1}{16}$  (f) 2, $2e^4$
(g) 1, $\tfrac{1}{3}$  (h) 0, $\tfrac{2}{5}$  2(a) 1, $\tfrac{1}{2}$  (b) 1, 4  (c) 0, $-\sqrt{3}$
(d) $\sqrt{3}, -\sqrt{3}$  (e) −2, ∞  3(b) $y = 5x - 1$  (c) $y = 2x + 3$
(d) $y = 1$  (e) $2x + 25y = 7$  (f) $y = 2e^2 x - e^2$
(g) $2y = x + 2\log 2 - 1$  (h) $2y = x + \log 2 - 1$  4(b) $x + 5y = 21$
(c) $x + 2y = 11$  (d) $x = 1$  (e) $10y = 125x - 123$
(f) $x + 2e^2 y = 1 + 2e^4$  (g) $y + 2x = \log 2 + 2$  (h) $4x + 2y = \log 2 + 4$
5(a) (2, 5)  (b) $\left(\dfrac{6}{7}, \dfrac{90}{7}\right)$  (c) (1, 5)  (d) $\left(-\dfrac{5}{4}, 24\right)$

## 14.3

1(a) (2, 4) max.  (b) (3, −6) min.  (c) (2, −16) min., (−2, 16) max.
(d) (−1, 6) max., (3, −26) min.  (e) (−3, 91) max., (2, −34) min.
(f) (0, 0) max., (−2, −16) min., (2, −16) min.  2(c) (0, 0)

(d) (1, −8)  (e) $(-\tfrac{1}{2}, 28\tfrac{1}{2})$  3(a) $\left(\dfrac{5}{6}\pi, 1\right)$ max.,

$\left(\dfrac{11}{6}\pi, -1\right)$ min.,  (b) $\left(\dfrac{1}{6}\pi, -1\right)$ min., $\left(\dfrac{1}{2}\pi, 1\right)$ max.

(c) $\left(\frac{2\pi}{3}, \frac{1}{2}\right)$ max., $\left(\frac{8\pi}{3}, -\frac{1}{2}\right)$ min.    4(a) (1, 0) max., (3, 4) min.
(b) (2, 12) min.    (c) (−1, −1) max., (4, 4) min.
(d) $(\frac{1}{3}\pi, \frac{1}{3}\pi - \sqrt{3})$ min., $(\frac{5}{3}\pi, \frac{5}{3}\pi + \sqrt{3})$ max.
(e) $(\frac{1}{3}\pi, \frac{4}{3}\pi - \sqrt{3})$ max., $(\frac{2}{3}\pi, \frac{8}{3}\pi + \sqrt{3})$ min.

5(a) $\left(-1, -\frac{1}{e}\right)$ min.    (b) $\left(\frac{3}{4}\pi, \frac{1}{\sqrt{2}}e^{3/4\pi}\right)$ max., $\left(\frac{7}{4}\pi, -\frac{1}{\sqrt{2}}e^{7\pi/4}\right)$ min

(c) $\left(\frac{1}{e}, -\frac{1}{e}\right)$ min.

## 14.4

**1** 30 m by 30 m    **2** 25 m by 50 m    **3** 3.9 cm    **4** 2 revs/sec    **5** points on x-axis are $x = 0.845$ and $x = 3.155$    **6** $\sqrt{\dfrac{a}{c}}$    **7** $r = \frac{1}{2}l$    **10** 60°
**11** 10 cm

## 14.5

**1** $2\pi r\,\delta r;\ 2\pi r\,\delta r + \pi(\delta r)^2$    **2** $4\pi r^2\,\delta r$    **3** 2 cm²    **4** 0.94 cm²
**5** 2500 cm³    **6** 1.2 cm³    **7** −0.40 cm³    **8** 0.54 cm    **9** 3.68 g
**10** $\dfrac{2}{5\pi} = 0.1273$    **11** $7.2\pi$ cm²/min    **12** $\dfrac{1}{\pi} \approx 0.318$ cm/sec

## 14.6

**1**(a) 40, 90    (b) 10, 10.    **2**(a) 80 ft/sec    (b) 16 ft/sec    (c) −80 ft/sec (opposite direction)    (d) after $2\frac{1}{2}$ sec    (e) −32 ft/sec²    (f) 80 ft/sec
**3**(a) 0, 288 cm/sec    (b) 0, 50 sec    (c) 150 cm/sec², 138 cm/sec²
(d) 25 sec.    **4**(a) 10 sec, 300 cm    (b) 5 sec    (c) no
**5**(a) 0, 32 cm/sec    (b) 0, $3\frac{1}{3}$ sec    (c) 40 cm/sec², −8 cm/sec²
(d) $1\frac{2}{3}$ sec    **6**(a) 10 cm    (b) 5 sec, $6\frac{2}{3}$ sec    (c) $5\frac{5}{6}$ cm/sec
(d) $5\frac{5}{6}$ sec    **7**(a) 9 cm, 8 cm    (b) $v = 3\cos t - 4\sin t$
(c) $a = -3\sin t - 4\cos t$    (d) $a + s = 5$

## 14.7

**1**(a) (2, 6)    (b) (7, 1)    (c) $(3\frac{3}{4}, 3\frac{3}{4})$    (d) (14, 8)
**2** (0, −1) max, $(\frac{1}{3}\pi, -1\frac{1}{2})$ min, $(\pi, 3)$ max., $(\frac{5}{3}\pi, -1\frac{1}{2})$ min., $(2\pi, -1)$ max.
**3** $\dfrac{1 - 2x - y}{2y + x + 1}$; (2, 0), −1; (−1, 0), ∞; (0, 1), 0; (0, −2), −1
**4** (0, 3) min., $(\frac{1}{2}\pi, 4)$ max, $(\pi, 3)$ min., $(\frac{3}{2}\pi, 4)$ max.,
$(2\pi, 3)$ min., points of inflexion at $x = \frac{1}{4}\pi, \frac{3}{4}\pi, \frac{5}{4}\pi, \frac{7}{4}\pi$, $y = 3\frac{1}{2}$ in each case
**5** $p < 0$; (0, 0) max., $(-\frac{1}{2}, -\frac{1}{8})$ min., $(\frac{1}{2}, -\frac{1}{8})$ min.    **6** (−1, 3) min.,
(1, 3) min.    **8** 30 gm cm⁻² sec⁻¹    **9** 0.234a    **10** $\dfrac{5}{8\pi}$ cm/sec

## 15.2

To save space arbitrary constants have been omitted.

**1**(a) $\dfrac{x^6}{6}$ (b) $-\tfrac{1}{2}x^{-2}$ (c) $\tfrac{3}{4}x^4$ (d) $-\dfrac{2}{x}$ (e) $\tfrac{1}{3}x^{1\tfrac{1}{2}}$ (f) $\tfrac{2}{3}x\sqrt{x}$

(g) $3x^{1/3}$ (h) $-\dfrac{2}{\sqrt{x}}$ **2**(a) $\tfrac{1}{3}x^3 + \tfrac{3}{2}x^2 + 5x$ (b) $\tfrac{1}{5}x^5 + x - \dfrac{2}{x} - \dfrac{1}{x^2}$

(c) $\tfrac{2}{3}x\sqrt{x} + 2\sqrt{x}$ (d) $\tfrac{1}{3}x^3 - \tfrac{3}{2}x^2 + 2x$ (e) $\tfrac{1}{5}x^5 - 2x^3 + 9x$

(f) $\tfrac{1}{3}x^3 + 2x - \dfrac{1}{x}$ (g) $\tfrac{1}{5}x^5 + x^2 - \dfrac{1}{x}$ **3**(a) 243 (b) $\tfrac{3}{4}$ (c) $34\tfrac{5}{6}$

(d) 33 (e) $4\tfrac{2}{3}$ **4**(a) 4 (b) $13\tfrac{1}{3}$ (c) $9\tfrac{5}{8}$ (d) $8 + \log 3$ (e) 72

**5**(a) $\tfrac{1}{3}\sin 3x$ (b) $-\tfrac{1}{2}\cos 2x$ (c) $-\cos(x + \tfrac{1}{6}\pi)$ (d) $\sin(x - \tfrac{1}{3}\pi)$

(e) $\tfrac{1}{4}\tan 4x$ (f) $\tfrac{1}{2}\tan(2x - \tfrac{1}{6}\pi)$ **6**(a) $\tfrac{1}{2}$ (b) $\dfrac{\sqrt{3}}{2}$ (c) $\tfrac{2}{3}\sqrt{3}$

(d) $\sqrt{3} - 1$ (e) $\tfrac{1}{3}$ (f) $\dfrac{2}{\sqrt{3}}$ **7**(a) $6e^t$ (b) $e^{t-1}$

(c) $\tfrac{1}{2}(e^2 - 1)$ (d) $e - e^{-2}$ **8**(a) $\tfrac{1}{2}\log x$ (b) $\log(x + 1)$

(c) $\tfrac{1}{2}\log\tfrac{5}{3}$ (d) $2\log 2$

## 15.3

### Exercise A

**1**(a) $\tfrac{1}{2}x^2 + x + \log(x + 1)$ (b) $\tfrac{1}{2}x^2 - x + 3\log(x + 3)$

(c) $\tfrac{1}{2}x^2 + 2x + \log(x^2 + x + 1)$ **2**(a) $\tfrac{1}{8}(2\sin 2x - \sin 4x)$

(b) $\tfrac{1}{6}\sin 3x + \tfrac{1}{10}\sin 5x$ (c) $-\tfrac{1}{6}\cos 3x - \tfrac{1}{2}\cos x$ (d) $\tfrac{1}{4}$ (e) 0

(f) $\dfrac{\sqrt{3}}{8}$ **3**(a) 4 (b) $\sqrt{3} - 1$ (c) $\tfrac{1}{4}\sqrt{3}$ (d) $\tan x$

(e) $\tfrac{1}{2}x - \tfrac{1}{4}\sin 2x$ (f) $\tan x - x$ **4**(a) $\log(x + 2)^2 (x + 1)$

(b) $\log(x + 1)^2 \sqrt{2x - 1}$ (c) $\log\dfrac{x - 3}{2x + 3}$ (d) $\log\dfrac{x^3}{(2x - 3)^2}$

(e) $\tfrac{1}{3}\log\dfrac{(2x - 1)}{x + 1}$ (f) $\tfrac{2}{5}\log\dfrac{x - 1}{2x + 3}$ **5**(a) $\log\dfrac{5}{3}$ (b) $\log\dfrac{288}{49}$

(c) $\tfrac{1}{3}\log\dfrac{8}{5}$ (d) $\tfrac{1}{5}\log\dfrac{343}{54}$ (e) $\tfrac{1}{7}\log\dfrac{12}{5}$ (f) $\tfrac{1}{2}\log\dfrac{112}{9}$

### Exercise B

**1**(a) $\tfrac{1}{2}x^2 + 3x + \log(x + 2)$ (b) $\tfrac{1}{2}x^2 - 2x - \log(x - 2)$

(c) $\tfrac{1}{2}x^2 + x + \log(x^2 - x + 1)$ **2**(a) $-\dfrac{1}{12}\cos 6x + \tfrac{1}{4}\cos 2x$

(b) $\tfrac{1}{10}\sin 5x + \tfrac{1}{2}\sin x$ (c) $\tfrac{1}{6}\sin 3x - \dfrac{1}{10}\sin 5x$ (d) $\dfrac{\sqrt{3}}{4}$ (e) $\dfrac{\sqrt{3}}{8}$

(f) $-\tfrac{1}{3}$ **3**(a) $\tfrac{1}{2}$ (b) $\sqrt{3}$ (c) $\tfrac{1}{4}\sqrt{3}$ (d) $-\tfrac{1}{8}\cos 4x$

(e) $\tfrac{1}{2}x + \tfrac{1}{4}\sin 2x$ (f) $-\cot x - x$ **4**(a) $\log\dfrac{(x - 1)^2}{x + 2}$

(b) $\frac{2}{3}\log(3x-2)+\log(x-1)$  (c) $\log\dfrac{2x-5}{x+5}$  (d) $\log\sqrt{x}(x-5)^3$
(e) $\frac{1}{5}\log\dfrac{x-2}{x+3}$  (f) $\log\dfrac{2x-1}{3x-1}$  5(a) $\log 3$  (b) $\frac{1}{7}\log\frac{12}{5}$
(c) $-\log 3+\frac{1}{3}\log\frac{11}{8}$  (d) $\log\frac{9}{8}$  (e) $\frac{1}{3}\log\frac{5}{2}$  (f) $\log\frac{7}{8}$

## 15.4

**Exercise A**

1(a) $\dfrac{1}{6}(x^2+2)^6$  (b) $\dfrac{1}{7}(\sin x)^7$  (c) $e^{x^2}$  (d) $\log(x^2-1)$  (e) $\dfrac{1}{3}\tan^3 x$
(f) $\log \sin x$  (g) $\dfrac{1}{3}(x^2+4)^{1\frac{1}{2}}$  (h) $e^{\sin x}$  (i) $\log \tan x$  (j) $\dfrac{1}{6}\cos^3 2x$

2(a) $-\dfrac{1}{4(2x-1)^2}$  (b) $2\sqrt{1+x}$  (c) $u+\frac{1}{2}\log\dfrac{u-1}{u+1}$ where
$u=\sqrt{1+x^2}$  (d) $\frac{1}{2}\log(e^{2x}+1)$  (e) $\frac{1}{2}x\sqrt{1-x^2}+\frac{1}{2}\sin^{-1}x$
(f) $2\sqrt{x}-\log(x+1)-2\tan^{-1}\sqrt{x}$  (g) $\log \tan\frac{1}{2}\theta$  (h) $2\sin\sqrt{x}$

3(a) $-\dfrac{1}{x+2}+\dfrac{2}{(x+2)^2}-\dfrac{4}{3(x+2)^3}$  (b) $2\sqrt{x}-2\log(\sqrt{x}+1)$
(c) $-\dfrac{2}{1+\tan\frac{1}{2}\theta}$  (d) $e^x-\log(e^x+1)$  (e) $\log\dfrac{\sqrt{x+1}-1}{\sqrt{x+1}+1}$
(f) $\frac{1}{5}\log\dfrac{2\tan\frac{1}{2}\theta-3}{3\tan\frac{1}{2}\theta-1}$  (g) $\sqrt{x^2+1}$  (h) $\frac{1}{3}(1-x^2)^{1\frac{1}{2}}-(1-x^2)^{1/2}$
(i) $\log\dfrac{\sqrt{x-1}}{\sqrt{x+1}}$  4(a) $14\frac{2}{3}$  (b) $4\log\frac{3}{2}-1$  (c) $1+\frac{1}{2}\pi$  (d) $\frac{1}{4}\log 3$
(e) $\frac{1}{2}\log\dfrac{(e+1)^2}{e^2+1}$  (f) $\frac{1}{2}\log 2$

**Exercise B**

1(a) $\dfrac{1}{4}(x^3-1)^4$  (b) $-\dfrac{1}{5}\cos^5 x$  (c) $-e^{-x^2}$  (d) $\log(x^3-1)$
(e) $-\frac{1}{2}\cot^2 x$  (f) $-\log\cos x$  2(a) $\log(e^x+1)$  (b) $\tan\frac{1}{2}\theta$
(c) $\dfrac{x}{\sqrt{x^2+1}}$  3(a) $e^x+x$  (b) $\log\tan(\frac{1}{2}\tan^{-1}x)$
(c) $\dfrac{1}{5}\log\dfrac{2+\tan\frac{1}{2}\theta}{1-2\tan\frac{1}{2}\theta}$  (d) $2\log(\sqrt{x}+1)$
(e) $-\cos\theta+\frac{2}{3}\cos^3\theta-\frac{1}{5}\cos^5\theta$  (f) $-\frac{1}{2}\cot\theta$
4(a) $\sqrt{3}-1$  (b) 2  (c) $\frac{1}{6}\pi$

## 15.5

**Exercise A**
1 $x\sin x+\cos x$  2 $xe^x-e^x$  3 $\frac{1}{4}\sin 2x-\frac{1}{2}x\cos 2x$

143

4 $\frac{1}{2}x^2 \log x - \frac{1}{4}x^2$  5 1  6 $1 - 2e^{-1}$
7 $\frac{\pi}{18} + \frac{1}{9}$  8 $\frac{1}{2}(1 - \log 2)$  9 $2x \sin x + (2 - x^2)\cos x$
10 $\frac{1}{9}x^3(3\log x - 1)$  11 $e^x(x^2 - 2x + 2)$  12 $\frac{1}{x}(\log x + 1)$
13 $x(\log x - 1)$  14 $x \sin^{-1} x + \sqrt{1 - x^2}$  15 $e^x(x^3 - 3x^2 + 6x - 6)$

### Exercise B
1 $\sin x - x \cos x$  2 $-e^{-x}(x + 1)$  3 $\frac{1}{2}x \sin 2x + \frac{1}{4} \cos 2x$
4 $\frac{1}{4}e^{2x}(2x - 1)$  5 $\frac{1}{2}\pi - 1$  6 $e^2$  7 $\frac{1}{4}\pi - \frac{1}{2}\log 2$  8 $8\log 4 - 3\frac{3}{4}$
9 $x^2 \sin x + 2x \cos x - 2 \sin x$  10 $-2x^2 \cos \frac{1}{2}x + 8x \sin \frac{1}{2}x + 16 \cos \frac{1}{2}x$
11 $-\frac{1}{4}e^{-2x}(2x^2 + 2x + 1)$  12 $\frac{1}{4}x^4 \log x - \frac{x^4}{16}$
13 $2x(\log x - 1)$  14 $x \tan^{-1} x - \frac{1}{2}\log(1 + x^2)$  15 $\frac{1}{2}e^{x^2}(x^2 - 1)$


## 15.6

### Exercise A
1 $x^3 = 3y^2 + C$  2 $2x^3 + 3x^2 = 3y^2 + C$  3 $\sin x + \cos y = C$
4 $\tan x = y^2 + C$  5 $y^2 + y = x^2 + x + C$  6 $y^2 + y = \sqrt{x^2 - 1} + C$
7 $y^2 + 2y = 2e^x + C$  8 $(x^2 + 1)^{1\frac{1}{2}} = 3 \log y + C$  9 $y^2 = \log A(1 + x^2)$
10 $\sin x - \cos x = \log y + C$  11 $2 \tan y = e^{2x} + C$
12 $y^2 = \sin^{-1} x + C$  13 $2(x + 1)^3 = 3y^2 + 6y - 7$  14 $\sin x = \tan y - \frac{1}{2}$
15 $\log(y - 1) = \log 2(x - 1)$

### Exercise B
1 $2x^2 + 4x = y^4 + C$  2 $2(y + 1)^3 = 3x^2 + C$  3 $\sin 2y + 2 \cos x = C$
4 $x^3 + \cot y = C$  5 $(y + 1)^3 = x^3 + C$  6 $y^3 = (x^2 + 9)^{1\frac{1}{2}} + C$
7 $x^3 + 3x + 3e^{-y} = C$  8 $(y^2 + 4)^2 = 4 \log x + C$
9 $\log(y^2 - 1) = x^3 + C$  10 $e^y = \sin x + \tan x + C$
11 $\log y = xe^x - e^x + C$  12 $\log y = \tan^{-1} x + C$
13 $y^3 + \dfrac{1}{x + 1} - 2 = 0$  14 $\sin y + \cos 2x = 0$  15 $xy = 4(x - 1)$
16 $\sec y = 2 \sin x$


## 15.7

1 $\log(2x - 1)$  2 $x + \frac{1}{2}\log(2x - 1)$  3 $2\sqrt{2x - 1}$  4 $10\frac{2}{3}$  5 $\dfrac{128}{15}$

6 $-\frac{1}{3}(4 - x^2)^{1\frac{1}{2}}$  7 $1 + \frac{1}{2}\pi$  8 $\frac{1}{3}\pi - \dfrac{\sqrt{3}}{4}$  9 $\frac{3}{8}(x^2 + 1)^{4/3}$

10 $\frac{1}{2}x^2 + x + \tan^{-1} x$  11 $\dfrac{1}{8}x - \dfrac{1}{64}\sin 8x$  12 $\log(e^x + 1)$

13 $\log \dfrac{729}{64}$  14 $\dfrac{4}{7}$  15 $\frac{1}{4}\sin^4 x$  16 $\frac{1}{2}\log(x^2 + 1) - \tan^{-1} x$

17 $\frac{1}{2}x^2 + 2x + 2\log(x-1)$   18 $\frac{1}{2}\sin x + \frac{1}{14}\sin 7x$   19 2

20 $\frac{1}{4}x\sin 4x + \frac{1}{16}\cos 4x$   21 $-\frac{1}{14}\cos 7x + \frac{1}{2}\cos x$   22 $\log(e^{2x} + e^x)$

23 $-\frac{1}{\sqrt{x^2+1}}$   24 $-\frac{1}{4}x\cos 2x + \frac{1}{8}\sin 2x$   25 $\frac{-2}{3(x+1)^{1\frac{1}{2}}}$

26 $-\frac{2}{5}\sqrt{3-5x}$   27 $\frac{2}{3}(3\sqrt{3} - \pi)$   28 $\log\frac{x-1}{x}$

29 $\frac{1}{20}(5\cos x - \cos 5x)$   30 $\frac{1}{2}\log(e^{2x} + 1) + \tan^{-1}e^x$

31 $\log(1 + \sqrt{2}) - \frac{1}{\sqrt{2}}$   32 $\tan^{-1}x + \tan^{-1}y = C$; $-1$

33 $v = \frac{g}{k}(1 - e^{-kt})$, $s = \frac{gt}{k} - \frac{g}{k^2}(1 - e^{-kt})$   34 $\sqrt{\frac{g}{k}}$   35 $xy = 36$

## 16.2

1(a) 8   (b) 12   (c) $-10\frac{2}{3}$   (d) $\frac{1}{4}$   (e) $\frac{2}{3}$   (f) $2\frac{2}{3}$   (g) $\frac{7}{12}$

2(a) 144   (b) $10\frac{2}{3}$   (c) $1\frac{1}{8}$   (d) $1\frac{1}{3}$   (e) $\frac{125}{96}$   (f) $\frac{1}{12}$   (g) $\frac{8}{3}$

3 27:37   4 $\sqrt{3} - \frac{\pi}{3}$   5 $\frac{e^7 - e}{2}$   6 1, $10\log 10 - 9$   7 25.6

8 (2, 6), (6, 2), $\frac{80}{3} - 12\log 3$   9 (1, 5), (5, 5), $21\frac{1}{3}$

10 $4 + \log 5$   12 $\frac{k}{0.41}\left\{\frac{1}{v_1^{0.41}} - \frac{1}{v_2^{0.41}}\right\}$

## 16.3

1(a) $446\frac{2}{3}\pi$   (b) $\frac{56}{15}\pi$   (c) $35\frac{1}{3}\pi$   (d) $5\frac{1}{4}\pi$   (e) $20\frac{5}{6}\pi$   (f) $\frac{7}{24}\pi$

(g) $\pi\left(\frac{\pi}{2} - 1\right)$   2(a) $34\frac{2}{15}\pi$   (b) $32\pi$   (c) $12.8\pi$   (d) $51.2\pi$

(e) $\pi\left(2 - \sqrt{3} + \frac{\pi}{3}\right)$   (f) $34\frac{2}{15}\pi$   (g) $\frac{32\sqrt{2}}{15}\pi$   3 $\frac{8}{3}\pi$   4 $\pi$

5 $156\frac{4}{105}\pi$   6 9 cm   7 $22\frac{2}{3}\pi$

## 16.4

1(a) $(1\frac{1}{7}, 3\frac{5}{7})$   (b) (2.4, 0)   (c) (0, 1.6)   (d) $(3, 3\frac{3}{5})$   (e) $(2\frac{2}{5}, 16\frac{3}{7})$

2(a) (3, 0)   (b) $(3\frac{3}{14}, 0)$   (c) $(0, 1\frac{2}{3})$   (d) $(-\frac{1}{2}, 0)$   (e) $x = 4\log 2$

3 $\left(\frac{20}{3\pi}, \frac{20}{3\pi}\right)$   4 on the axis distant $\frac{3}{4}h$ from the vertex   5(a) $\frac{256}{15}$

(b) $x = \frac{12}{7}$   (c) $21\frac{1}{3}\pi$, (d) $x = \frac{8}{5}$   6 (0.1, 2.5)   7 $\frac{3}{8}r$ from the centre

8(a) 18   (b) $x = \frac{9}{16}\pi$   (c) $32.4\pi$

## 16.5

**Exercise A**

1  13   2  1   3  $\dfrac{\pi(\sqrt{3}+1)}{24}$   4  $\tfrac{1}{2}$   5  $\dfrac{2}{\pi}\log 2$

**Exercise B**

1  $13\tfrac{1}{6}$   2  32   3  $\dfrac{2}{\pi}$   4  $\tfrac{1}{2}(e^4-e)$   5  $\tfrac{1}{4}\log 5$

## 16.6

**Exercise A**
1(a) 1.1484   (b) 15.69   (c) $-0.496$   2(a) 0.2231   (b) 0.785
(c) 0.685   3(a) 0.693   (b) 1.178   (c) 6.938

**Exercise B**
1(a) 24.67   (b) 0.57   (c) 3.0915   2(a) 0.785   (b) 0.6236   (c) 0.9522

## 16.7

1(a) 25.6   (b) $\bar{x} = 2\tfrac{6}{7}$   (c) $64\pi$   (d) $\bar{x} = 3.2$   (e) $54\tfrac{6}{7}\pi$   (f) $y = 5.6$
2  (0, 0.3536)   3(a) 84   (b) $85\tfrac{1}{3}$   (c) $85\tfrac{1}{3}$   4  $x = 5\tfrac{1}{3}$   5  180
8  $4 - 3\log 3,\ \tfrac{8}{3}\pi$